COLLEGE OF MARIN LIBRARY
KENTFIELD, CALIFORNIA

WITHDRAWN

D1602451

CIVIL DEFENSE:

A Choice of Disasters

CIVIL DEFENSE:

A Choice of Disasters

Edited by
**John Dowling
Evans M. Harrell**

American Institute of Physics
New York

This volume was prepared by a study group of the Forum on Physics and Society of The American Physical Society. The American Physical Society has neither reviewed nor approved this study.

Library of Congress Cataloging in Publication Data

Civil defense.

Filmography: p.
Includes bibliographies and index.
1. United States—Civil defense. 2. Civil defense. I. Dowling, John. II. Harrell, Evans. III. American Institute of Physics.
UA927.C47 1986   363.3'5'0973   86-28785
ISBN 0-88318-512-1
Copyright © 1987 by American Institute of Physics
All rights reserved.
Printed in the United States of America.

Contents

Preface ... vii
About the Authors ... xi
Acronyms and glossary ... xv

Chapters

1: **The setting for civil defense: An overview**
   Paul P. Craig ... 1
2: **History of American attitudes to civil defense**
   Spencer R. Weart ... 11
3: **FEMA: Programs, problems, and accomplishments**
   John Dowling ... 33
4: **Effects of nuclear weapons**
   L. C. Shepley ... 47
5: **Nuclear radiation and fallout**
   Cliff Castle ... 65
6: **Sheltering from a nuclear attack**
   James W. Ring ... 77
7: **Maintaining perceptions: Crisis relocation in the planning of nuclear war**
   John Hassard ... 85
8: **Civil defense in other countries**
   Evans M. Harrell ... 105
9: **Civil defense implications of nuclear winter**
   Barbara G. Levi ... 125
10: **Long-range recovery from nuclear war**
    Ruth H. Howes and Robert Ehrlich ... 139
11: **Political and psychological issues in civil defense**
    Robert Ehrlich and Ruth H. Howes ... 153

Appendices

A: **Miscellaneous technical aspects relating to civil defense**
   A.1. Lifeboat analogy to civil defense ... 163
   A.2. High-risk areas map ... 165

A.3.	Standard radiation decay model for fallout	166
A.4.	The "penalty" table	166
A.5.	Area dose-rates after a hypothetical attack	167
A.6.	Guidelines for shelter and operational activities	167
A.7.	Exposures in shelters of low protection factors	168
A.8.	Fallout conditions from a random assumed attack	168
A.9.	Comparison of relative costs for civil defense	169

B: Should we protect ourselves from nuclear weapon effects?
    Carsten M. Haaland   171

C: The U.S. needs civil defense
    Roger J. Sullivan   199

D: Under the mushroom cloud
    Barry Casper   203

E: Nuclear winter and the strategic defense initiative
    Caroline Herzenberg   207

F: Aids for teaching civil defense
    L. C. Shepley   215

G: Film bibliography
    John Dowling   223

Index   229

Preface

Civil defense in a nuclear war has been a controversial topic since the bombs were dropped on Hiroshima and Nagasaki. Today's arguments, pro and con, follow those made in the 1950s and early 1960s. Pro states: we need civil defense as a form of insurance against the very real possibility of nuclear war. Con replies: civil defense threatens the other side and thus increases the likelihood of nuclear war. Pro: protection against radiation and blast is well within engineering capabilities. Con: the expense is great, and uncertainties are so large that it is unlikely to be effective in practice. Pro: we need evacuation plans so that we can move people away from where the weapons are likely to hit. Con: you cannot get out of New York City at 5 p.m. on a work day. And so it goes.

Why is there so much opposition to civil defense when the purpose of the program is to save lives? The reasons for this opposition are varied and complex. We are dealing with the public's perception of the threat of nuclear war formed over the years and with what can be realistically done to meet this threat. Further, we need to determine if there are conditions that would enable the United States to implement a national civil defense program that is acceptable to our society.

The explosion of the Soviet A-bomb in 1949 and the start of the Korean War set off an exuberant wave of civil defense in the early 1950s. This wave ebbed and flowed until a year or two after the Cuban Missile Crisis when it just seemed to fade away. The schoolchildren of the 1950s now remember the civil defense drills, proposed dog tags for kids, and bomb and fallout shelters in their local neighborhoods. But government films with directions on how to build your own fallout shelter, and with Bert the Turtle teaching children how to "duck and cover" were made suspect by such Hollywood films as *Fail-Safe, Dr. Strangelove,* and *On the Beach.* The trade of New York for Moscow in *Fail-Safe,* the military running amuck in *Dr. Strangelove,* and the impossible radiation scenario in *On the Beach* were, respectively, implausible, greatly exaggerated, and scientifically flawed. While these artistic visions were not very realistic, they grasped something of the internal, relentless logic of the arms race, a logic which may be bringing the prospect of global annihilation closer to reality. A recent film, *The Atomic Cafe,* points up the whole civil defense controversy. An audience viewing this film greets the "duck and cover" clips from official government films with derisive laughter. Why such a dramatic change in the attitude of the general public over the years? We went from acceptance of backyard fallout shelters to outright opposition by some medical doctors who refused to participate in readying hospitals to accept casualties in a war. And officials of some communities even refuse to work on civil defense evacuation plans.

There is a sense of *déjà vu* when one reads the literature of the 1950s and 1960s on nuclear war and civil defense. The devastation of nuclear war, the civil defense problems of outspoken critics' opposition and public apathy, and how to implement population protection and evacuation were all discussed then. But the problems have intensified. Formerly, the relatively slow bombers allowed for many hours warning and could be countered by air defenses. The relatively small numbers of bombs that would have made it through on bombers, could largely have been countered by a well-planned and efficiently executed civil defense program. Today submarine and cruise missiles are minutes away. Each superpower has to face thousands of strategic nuclear warheads and both Europes face many more thousands of tactical nuclear warheads.

Worsening international tensions, the deployment of SS-20's, cruise and Pershing II missiles in Europe, the resurgence of cold war rhetoric, the political activism of the peace movement, and the media attention to these matters have given rise to a greater awareness of the arms race and the consequences of nuclear war. Misunderstanding still prevails. People are aware, for example, that civil defense in the United States has to do with fallout shelters and crisis relocation (now currently out of favor). Many have the impression that a large centralized bureaucracy in Washington has expended great quantities of money in order to draw up elaborate evacuation plans for each community in the country. In fact, the Federal Emergency Management Agency (FEMA) is as embattled an agency as Washington has ever seen. There was an attempt to revitalize FEMA with a $4.2 billion dollar budget to implement a comprehensive civil defense program over a seven-year period. But FEMA's requests for funds to start that program were rejected by Congress in the 1983 to 1985 fiscal years (FY), and the budget for FY 1986 is well below that of FY 1985. FEMA is now restricted primarily to training and coordination of agencies at the state level, and these agencies are generally responsible for drawing up relocation plans patterned on a small number of federally funded prototype studies. The actual plans are quite rudimentary. In short, FEMA always seems to be in perpetual turmoil as is evidenced by these budget crises and the furor over crisis relocation.

According to its Congressional mandate, FEMA is charged with the legal and moral responsibility of protecting U.S. citizens from nuclear attack and other disasters, natural or man-made. Many people also believe that civil defense affects the strategic balance between the U.S. and the U.S.S.R., while others argue that the level of destruction would be so great with or without civil defense that deterrence is not materially affected. Since the Carter administration, the enhancement of deterrence has been officially regarded as an important secondary goal of civil defense. The scenario of a possible nuclear attack provides a natural way to classify civil defense activities: long-range preparations in peacetime, action in a crisis, sheltering during an attack and in the short term afterwards, and long-term survival and recovery. The difficulties in each period are great, and few people, whether they support or oppose a strong civil defense program, doubt that a nuclear war could be anything other than the greatest disaster ever experienced, even if good civil defense plans are in effect. Owing to budget constraints most civil defense planning has concentrated on action in a crisis and during an attack. The problems considered deal with relocation, communications, transportation and engineering, and sheltering—all of which involve the physics and engineering of shielding against radiation and blast.

All aspects of the arms race affect civil defense: attack scenarios, hardening of targets, warhead yields, accuracy, MIRVing (multiple independently targetable reentry vehicles), the Strategic Defense Initiative (SDI), and the possibility of nuclear winter. Counterforce and countervalue attacks are attacks on military targets and on cities, respectively. Counterforce targets such as missile silos are hardened and require accurate, ground-burst weapons to ensure destruction. Ground-burst weapons produce large amounts of fallout. There has been no serious attempt to address the problem of colocation of strategic installations near population centers. Counterforce targets such as the submarine bases in Groton, Connecticut and Seattle, Washington or the military-industrial complex spread around Los Angeles would probably be attacked with air-burst weapons which would ensure widespread destruction of major cities. If the warhead yield is large the problem is exacerbated. The higher the accuracy, the lower the yield needed. MIRVing means more warheads to contend with, but they generally have lower yields. A successful SDI would ensure that few warheads would penetrate our defenses, and consequently, make civil defense much, much easier. A successful SDI would make a strong case for implementing a comprehensive civil defense program. Often the consideration of interaction between the arms race and civil defense has been left out of the debate.

The data on what would actually happen in a nuclear attack on human populations are rather sparse. The size of the weapons used and the tragic lessons learned at Hiroshima and Nagasaki do not extrapolate very well to a full-scale nuclear attack. Many of the hazards of nuclear war are not easily understood and involve more in terms of civil defense than the traditional protection from blast and radiation in shelters. The Office of Technology Assessment report states this best: "The effects of a nuclear war that cannot be calculated are at least as important as those for which calculations are attempted. Moreover, even these limited calculations are subject to very large uncertainties." The phenomenon of electromagnetic pulse can cause disruption of communication and equipment malfunction, but the exact nature and extent of the problem are not known. Some of the most important issues have to do with the aftermath of a nuclear war, when one has to hope for at least a minimal amount of human survival. Theoretical studies by Crutzen and Birks and by Turco et al. have indicated that even a limited nuclear war might cause long-term climatic and ecological changes, conceivably threatening the extinction of humankind. Others have said that such speculation is unsupportable. Obviously, there is a pressing need for better scientific understanding of these issues if we are to make intelligent decisions about how to deal with the threat of nuclear war, especially in terms of civil defense.

The Forum on Physics and Society of The American Physical Society has been working to achieve this better understanding. Our work was undertaken with the encouragement and support of the Forum, which we gratefully acknowledge. The final product is our own. It has not been reviewed or endorsed by The American Physical Society.

Our group consists of eleven physicists—all of whom are actively interested in education about nuclear war, although generally they are not professionally involved in arms control or weapons research. The group is composed of both opponents and advocates of civil defense, as comes through in some of the chapters, although most members have tempered their views during the course of this study. We started our project by searching through the vast literature on civil defense and then sorting out the important issues and papers. A rough form to the study began to take place and in June 1984 most of the group met for a work session in Washington, D.C. There we formed an outline of what issues we thought the book should cover and various people took responsibility for individual chapters. The following day we all went to a full-day briefing which FEMA had kindly arranged for us. The meeting at FEMA helped to broaden our viewpoints, and the group met together afterwards to assimilate what we had learned. The first drafts of the chapters were mostly completed during the autumn of 1984 and circulated to each member of the group for comments and criticism. Dowling and Harrell met in early March of 1985 to give a critical reading to all chapters. In April 1985 we all got together for a three-day work session at Mansfield University. Each chapter was examined critically by most of the eleven group members and more comments and suggestions for a final draft were made. In the summer of 1985 the two editors filled in the gaps and eliminated the information duplicated between chapters.

Every attempt was made to create as impartial a report as possible. This was not easy to achieve. When the first draft of this preface was circulated to our working group, some parts were said to be too soft on civil defense and others too hard. Constructive criticism and considerable good will on the part of the group members produced a preface which has attained, if not consensus, at least the consent of all. However, the same cannot be said for the whole report. Each chapter is the work of a particular author (or authors) and may contain material which reflects the author's individual viewpoints on particular issues.

Civil Defense: A Choice of Disasters is written for both interested laypersons and for those who wish an introduction to the broad range of topics comprising "civil defense." Most undergraduates should be able to understand each chapter thoroughly, except for parts of the more technical chapters. The book begins with an overview of civil defense, followed by histories of American attitudes towards civil defense and of FEMA and its predecessors. We examine the technical issues of the effects of weapons, radiation, and sheltering. The topics of crisis relocation and foreign civil defense are covered next, then nuclear winter and long-range recovery. The final chapter deals with the political issues associated with civil defense. There are several appendices which deal with the more technical aspects, three position papers on civil defense—two pro and one con which were given at a Forum symposium in Detroit in March 1984. There are also two appendices which will help educators who wish to incorporate civil defense issues into their courses. We hope the book will be useful both for courses dealing with the arms race and the threat of nuclear war and for the interested layperson.

John Dowling
Mansfield, PA

Evans M. Harrell
Atlanta, GA

About the authors

Cliff Castle is a physics teacher at Jefferson College, Hillsboro, MO. Since 1977 he has worked extensively as a Radiological Instructor in FEMA Region VII. He has been a member of the Missouri Nuclear Emergency Team since its inception in 1979. The technology of nuclear weapons and the history of science are his primary research interests.

Paul P. Craig is a Professor in the Department of Applied Science, University of California, Davis and faculty associate at the Lawrence Berkeley Laboratory. He has a B.S. in physics from Haverford College and a Ph.D. from Cal Tech in cryogenics. He did basic research (unclassified) at Los Alamos from 1959–1962, was at Brookhaven National Laboratory from 1962–1971, and was a Guggenheim Fellow in 1965–1966. He was at NSF from 1971–1975 and has been at the University of California since 1975. He was a member of The American Physical Society's Panel on Public Affairs in 1972–1973 and has served on the Board of Directors and the Executive Committee of the Environmental Defense Fund. He is currently Vice-Chairperson of the Forum on Physics and Society. He is coauthor of the text *Nuclear Arms Race: Technology and Society*.

John Dowling is Professor of Physics at Mansfield University, Mansfield, PA. He has a B.S. from the University of Dayton, and an M.S. and Ph.D. from Arizona State University. His primary research area is in arms control, specializing in ways to inform the public about the issues of the arms race. He was film review editor of the *American Journal of Physics* from 1976–1982 and of the *Bulletin of the Atomic Scientists* from 1979–1984. He edited the American Association of Physics Teachers book *The Cinescope of Physics*, authored the book *War-Peace Film Guide*, coauthored the *1984 National Directory of AV Resources on Nuclear War and the Arms Race*, and coauthored the AAPT's *Resource Letter: Physics and the Arms Race*. He has been the editor of the Forum's newsletter *Physics and Society* since 1980. During the 1985–1986 academic year he spent his sabbatical at Michigan State University. He hereby acknowledges the generosity of the MSU Department of Physics and Astronomy during his stay. In particular, he would like to thank Jack Bass who made the facilities of the department available to him.

Robert Ehrlich chairs the Physics Department at George Mason University, Fairfax, VA. He has held faculty positions at the University of Pennsylvania, Rutgers University, and the State University of New York, where he chaired the Physics Department at SUNY–New Paltz. Ehrlich's Ph.D. in physics from Columbia University in 1964 was in the area of experimental particle physics, an area of research accounting for a majority of his 30 publications. In recent years his published works have related to the effects of nuclear war and nuclear weapons policy. Dr. Ehrlich has written three books, the most recent of which, *Waging Nuclear Peace* (SUNY Press, Albany, NY, 1985), is intended for use as a text in interdisciplinary survey courses treating nuclear weapons policy and technical issues. He organized the Nuclear War Education Conference held in April 1986.

Evans M. Harrell is Associate Professor of Mathematics, Georgia Institute of Technology, specializing in mathematical physics and operator theory. He graduated with a B.S. in physics from Stanford in 1972. His Ph.D. was obtained from Princeton University in 1976. He has taught at Haverford College, University of Vienna, Massachusetts Institute of Technology, and Johns Hopkins University. He is author of numerous research articles and translator of four technical books (from German).

John Hassard is a high-energy physicist who worked for Harvard University at Cornell University's Electron Storage Ring in Ithaca. He did his Ph.D. for Manchester University, working in Hamburg, West Germany. In January 1986, he joined the Faculty of Imperial College, High Energy Physics, London, England SW7 2AZ and now commutes between London and Geneva, Switzerland. He was director of the Physicians for Social Responsibility's Task Force on Nuclear War Evacuation while at Cornell. Professor Hassard is also a member of the American Committee for East-West Accord, based in Washington, D.C.

Ruth H. Howes is a Professor of Physics at Ball State University in Muncie, IN. She holds a B.A. in physics from Mount Holyoke College and an M.S. and Ph.D. in low-energy nuclear physics from Columbia University. She has taught at the University of Oklahoma and Oklahoma City University. She has been teaching courses on nuclear issues since 1976. In 1984–1985 she was a Visiting Scholar at the Arms Control and Disarmament Agency in the Bureau of Verification and Intelligence. Her research interests also include the application of nuclear physics to problems in geology, archaeology, and art history.

Barbara G. Levi is a research physicist at the Center for Energy and Environmental Studies, Princeton University. She taught physics at Georgia Institute of Technology and at Fairleigh Dickinson University. Levi's research work at Princeton has been principally in the areas of energy and nuclear weapons policy. She also writes research news stories for *Physics Today* and has consulted for the Congressional Office of Technology on such studies as *Nuclear Proliferation and Safeguards*.

James W. Ring is Professor of Physics at Hamilton College in Clinton, NY. He obtained his Ph.D. in physics from the University of Rochester in 1958. He was at the Atomic Energy Research Establishment in Harwell, U.K., the Physical Chemistry Lab at Oxford University, U.K., and visiting fellow at Princeton's Center for Energy and Environmental Studies.

Lawrence C. Shepley is an Associate Professor in the Physics Department and Associate Director of the Center for Relativity at the University of Texas.

He received his B.A. from Swarthmore College and his M.A. and Ph.D. from Princeton University in 1965. He was a Post-Graduate Research Physicist at the University of California, Berkeley and went to the University of Texas, Austin in 1967. His research interests are in the theory of relativity, cosmology, and the foundations of mechanics. In addition to authoring research papers, he is the coauthor of *Homogeneous Relativistic Cosmologies* (Princeton University Press, Princeton, NJ, 1975). He coedited *Cosmology: Selected Reprints* and *Spacetime and Geometry—The Alfred Schild Lectures*.

Spencer R. Weart is the Director of the Center for History of Physics at the American Institute of Physics in New York. His B.A. is from Cornell University and his Ph.D. is from the University of Colorado. He did postdoctoral work in solar physics at the California Institute of Technology and in history at the University of California, Berkeley. He is currently working on a book on the history of nuclear imagery.

Acronyms and glossary

ABM	antiballistic missile
ACDA	United States Arms Control and Disarmament Agency
ALCM	air-launched cruise missile
alpha	an alpha particle consists of two neutrons and two protons bound together; an alpha particle can be formed when a nucleus undergoes radioactive decay; alpha particles are usually absorbed by a few centimeters of air
ASAT	antisatellite, usually used in context of ASAT weapons
bar	metric unit of pressure (1 bar = 0.987 atmospheres = 100 000 Pascals)
beta	a beta particle is a positive or negative electron and can be formed when an atom undergoes radioactive decay; it can usually be absorbed by several sheets of paper
BMD	ballistic missile defense—measures for defending against incoming enemy missiles; SDI would be the latest version of BMD
BSPP	Broadcast Station Protection Program—a FEMA program designed to protect radio and TV stations in the event of a nuclear attack
CD	civil defense
CIA	Central Intelligence Agency
cm	centimeter (2.54 cm = 1 in.)
counterforce	an attack on military targets
countervalue	an attack on civilian targets, such as cities
CRP	crisis relocation plan
DCPA	Defense Civil Preparedness Agency
decibar	unit of pressure, 1 decibar = 10 000 Pascal
EMP	electromagnetic pulse—in a nuclear explosion a small part of the energy released goes into creating a strong electromagnetic field (similar to a lightning bolt, but with higher voltages and occurring in a shorter time); EMP has the potential to knock out unprotected electrical circuits over a large area underneath a nuclear explosion
EOC	Emergency Operations Center—a FEMA program that provides facilities which will act as command centers to direct recovery and rescue operations in the event of a nuclear attack

EPZ	emergency planning zone
eV	electron volt—a metric unit of energy; an electron gains 1 eV when it is accelerated through an electrical potential difference of one volt, $1\text{ eV} = 1.6 \times 10^{-19}$ Joules
FCDA	Federal Civil Defense Act (or Administration or Agency)
FEMA	Federal Emergency Management Agency
FY	fiscal year
gamma	gamma particles are high-energy photons (light waves); gamma radiation can occur when a nucleus undergoes radioactive decay; it can often penetrate several centimeters of lead and thus poses a high danger to life forms
GAO	General Accounting Office
GLCM	ground-launched cruise missiles
GNP	gross national product
gray	unit of radiation (one gray = 100 rads, see below)
ICBM	intercontinental ballistic missile
IEMS	Integrated Emergency Management System
INR	initial nuclear radiation—a nuclear explosion produces energy in three forms: light, blast, and initial nuclear radiation; this is to be distinguished from fallout radiation which results when the radioactive particles (the INR) produced in the explosion are deposited on the material which the fireball contacts when it touches the earth
Joule	metric unit of energy, equivalent to 0.000 239 food-type calories
keV	thousand electron volts (see eV)
km	kilometer (1.6 km = 1 mile)
kPa	kilopascal, unit of pressure
kt	kiloton (equal to 10^9 food-type calories)—a measure of the energy released in a nuclear explosion; the Hiroshima bomb was approximately 15 kt
MeV	million electron volts (see eV)
MIRV	multiple independently targetable reentry vehicle
Mt	megaton (equal to 10^{12} food-type calories)—a measure of the energy released in a nuclear explosion; the Hiroshima bomb was approximately 0.015 Mt
N	Newton, a metric unit of force
NACP	National Attack Civil Preparedness
NAS	National Academy of Sciences
NCP	Nuclear Civil Protection

Acronyms and glossary xvii

neutron	an uncharged particle which together with protons constitute the primary building blocks of the nucleus of an atom; neutrons are part of the INR released by a nuclear explosion and have important radiological effects on life forms
NSDD	National Security Decision Directive
NTIS	National Technical Information Service
OTA	Office of Technology Assessment
pascal	metric unit of pressure (1 atmosphere = 101.3 kPa = 14.7 psi)
PF	protection factor—a measure of how much the incoming radiation is reduced by matter interposed between the radiation source and the eventual receiver, a protection factor of 10 reduces the radiation by a factor of 10
psi	pounds per square inch (1 atmosphere = 14.7 psi, 1 psi = 6.9 kPa)
PSR	Physicians for Social Responsibility
rad	radiation absorbed dose—one rad is that amount of radiation which deposits 0.01 Joule of energy into one kilogram of absorbing material
radioactive decay	the process by which a nucleus spontaneously breaks up into two or more particles and emits either an alpha, beta, or gamma particle or a neutron (or some combination thereof); when these emitted particles hit living tissue they can damage the molecules that make up the tissue
R&D	research and development
RDO	radiological defense officer
rem	roentgen equivalent man (or mammal), measurement of radiation dose received; typically the average U.S. citizen receives a dose of 0.2 rem per year
Roentgen	one Roentgen is that amount of radiation that deposits 0.008 76 Joules of energy into one kilogram of air
SAC	Strategic Air Command
SALT	Strategic Arms Limitation Treaty (SALT I, the ABM treaty ratified in 1972, SALT II signed in 1979, but never ratified by the U.S.)
SDI	Strategic Defense Initiative, popularly known as Star Wars
SF	Swiss francs
SLBM	submarine-launched ballistic missiles
START	Strategic Arms Reduction Talks, President Reagan's follow-up to the SALT talks
terragram	10^{12} grams

Chapter 1
The setting for civil defense: An overview

Paul P. Craig

The setting

The threat of nuclear war requires that civil defense deal with two broad problems: (1) protecting the lives of civilians against any possible attack and (2) minimizing the damage to and ensuring the recovery of the industrial, agricultural, economic, and social infrastructure of the nation. While there is no explicit connection between civil defense efforts and the military, there is obviously an interaction between the two. For example, the location of important military targets and manufacturing facilities, different attack scenarios that could occur, and the types and sizes of weapons used all have an impact on the size and the emphases of the appropriate civil defense measures that must be taken. In addition, civil defense would have some impact on the strategic balance. If one side evacuates its cities does this mean an attack is likely to ensue? Civil defense cannot be viewed as simply protecting the civilian population; it goes far beyond this.

In this book we examine many of the considerations which must be taken into account to understand the potential and the limits of civil defense in a nuclear age. The authors of this study have backgrounds in physics and the physical sciences, and thus our attention is focused on the physical and biophysical implications of civil defense. However, the problems involved in assessing civil defense involve virtually every aspect of society. In this introduction we raise a spectrum of issues associated with civil defense in a nuclear age. Cross references are included to particular chapters where most of the points are discussed in detail, and where citations to specific references in U.S. and foreign literature may be found.

Any discussion relating to nuclear war and the preparation for nuclear war must take into account the enormous uncertainty which surrounds virtually every aspect of the matter. In a few areas knowledge is relatively solid; for example, the blast pressure associated with the explosion of a nuclear weapon of known size at a particular altitude can be determined for any location. However, the defense can never know in advance the size or number of weapons targeted on a particular location, or the height of burst, or the weather. The interaction between civil defense and the probability of war remains almost totally opaque. The effects of "nuclear winter" have been the subject of much speculation, but the area is fraught with uncertainty. Our approach to discussing the role of civil defense in the nation's nuclear strategy places emphasis not only on that which is known, but also on the many other considerations which are clearly important, but about which little is known.

Literature on the ethics of war[1] emphasizes the doctrine of *proportionality*—the view that a military action undertaken should be scaled in proportion to the required ends. The letter of the American Catholic Bishops in 1983 [2] takes this perspective in an analysis which supports the viewpoint that nuclear weapons can legitimately be a part of the arsenals of moral nations provided that there is a process in motion which will (in all likelihood) move toward their eventual elimination. This view is in accord with the way in which many wars—though by no means all—were conducted in previous times. There was a reasonably clear-cut distinction between individuals who were part of the military (be they professional soldiers, volunteers, conscripts, or hired mercenaries) and civilians.

An example may help illustrate the historic view and the change. The Battle of Bull Run was announced in the Washington newspapers, and many residents of that city took the trouble to go to the site of the prospective battle in order to have a firsthand look at the events of the day. Certainly these revellers were aware that an inadvertent misfire might cause some grief. Yet they were not overly concerned that they might themselves be caught up in the conflict. Similarly, at the Battle of Gettysburg, where (it is recorded in exhibits on display at the Gettysburg Museum) the farmers were advised to, and did, go home during the battle, but shortly thereafter returned to watch over their fields. Several instances are recorded of citizens caught up in the battle who appealed to the military to check their status out with the folks in town, and who, following investigation, were sent home.

By the end of the Civil War, things changed. General William Tecumseh Sherman's "March to the Sea" was one of the more brutal episodes in the history of warfare.[3]

Suffering, to be sure, occurred in earlier times. *Mother Courage*, one of Bertolt Brecht's more impressive plays, took as its focus the starvation and disruption experienced by the peasants during the Thirty Years War. Yet, even here the peasants were not actively a part of the war, but were rather the victims of ravages due to hungry troops finding their meals where they could.

To the extent that a historical consensus on noninvolvement of civilians existed, it evaporated in the twentieth century. The collapse was partially visible in World War I; by World War II it was clear. By then the role of civilians in warfare had become significant, and the strategies of Adolf Hitler took clear account of the role of civilians as conscript labor to make the war machine function.

The Nazi attacks on Great Britain remain possibly the most vivid reminder of the importance of attacks on civilian populations. Hitler had a clear policy of destroying the industrial might of Great Britain, and attempted to accomplish this by a series of coordinated raids on major cities.

Civil defense played a key role in permitting Britain to withstand these attacks. With ample warning provided by observers and by radar, citizens disappeared into the safe environment provided by bomb shelters beneath their buildings. They even developed a sense of camaraderie in the shelters which is especially well described in *Weapons and Hope*.[4] Toward the end of World War II the Germans used the V-2, an early version of a guided missile. These weapons traveled so fast that little warning was possible. Had they been operational during the Battle of Britain, the role of bomb shelters might have been considerably reduced.

The allies also used attacks against civilians as a part of their World War II strategy. The firestorms following the bombings of Dresden (for a fictionalized account, see Vonnegut[5]) and Tokyo are examples. The U.S. nuclear attacks on Hiroshima and Nagasaki were specifically intended to produce mass destruction in order to destroy Japanese morale.

What were the key attributes of the British civil defense programs in World War II? First, the Germans possessed weapons capable at best of destroying but a block or so of land area. Second, the total number of incoming bombers in any one raid was small enough that only a relatively small number of injuries per raid was anticipated. Third, though the Royal Air Force was able to down only about 10% of the incoming planes, by the time three or four raids were mounted by the Nazi's, only 65% or so of the initial planes were still flying. An attrition rate of 10% per raid was quite enough to effectively demolish the Nazi attack, and to allow England to finally win the Battle of Britain.

Phases of defense

These historical observations provide some guidance for characterizing the potential role of a civil defense program in a nuclear age. To do so it is useful to review the kinds of situations with which a civil defense system might be concerned.

I begin by subdividing the possible spectra of events into phases. These range from events leading up to nuclear war through postattack recovery. We list the phases here and discuss each in more detail later on in this chapter.

Prewar. The United States is presently in a period of tension and uncertainty. Military expenditures are high and are growing. A nuclear attack could come with no explicit prior notice, or it could follow a short or long period of escalating international tensions, including non-nuclear conflict between the U.S. and/or its allies and the Soviet Union and/or its allies. In a time of heightened tension crisis relocation has been proposed (see Chaps. 3 and 7). Planning of types of civil defense must take into account the possibility of long periods of tension. Thus plans which require, for example, evacuation of cities cannot often be put into effect. Nor would it be easy to maintain the population of cities for long periods in distant locations.

Initial attack. Many different kinds of attack can be envisaged. These range from an attack of just a few weapons from a known or unknown adversary, through a continuing war in which nuclear weapons arrive over a period of weeks or months, to a massive spasm in which most of the nuclear weapons in the arsenals of the U.S. and the U.S.S.R. are detonated in a short period (see Chap. 4).

Postattack. Before, during, or immediately after an attack, those citizens who can will seek protection in blast and/or fallout shelters. The time period over which sheltering is required may last from days to months. This is the time during which fallout decays to a level permitting persons to spend significant amounts of time outside of shelters. The immediate postattack environment is the one with which most civil defense planning has been concerned (see Chap. 3).

Postattack recovery. As survivors emerge from shelters they will need to find food, and begin the recovery of society. Their ability to do this will depend on the magnitude of the attack, on their physical, physiological, and psychological states, the degree to which needed resources have survived (including food and means of reconstructing the technological infrastructure), and the possibly changed climate into which they will emerge. Each of these considerations is complex and uncertain. For example, if a "nuclear winter" situation has developed (see Chap. 9), the survivors may emerge into a bitterly cold, dark world in which there is little food and little possibility to produce more.

Attack scenarios

These phases of defense can be considered in the context of possible attacks. There are an enormous number of ways in which the United States could be attacked. The imagination of the reader will certainly be adequate to produce many scenarios. For our present purposes it is only necessary to indicate ranges of possibilities and the extremely diverse set of impacts that different attacks would have on the United States.

Perhaps the most frequently discussed attack is all-out war. This was the kind of situation dealt with in the TV movie shown in 1983, *The Day After*, in which virtually every weapon in the Soviet arsenal was apparently set off. In *The Day After*, the United States was virtually destroyed. Nuclear winter might well follow such an attack.

At an opposite extreme is a situation in which restricted nuclear warfare takes place. One example might be an accidental launching of a Soviet missile which cannot be recalled. The target city would have warning of a few minutes at most. However, after the explosion, resources would be available from the outside to assist in recovery. Nuclear winter would not be an issue.

An intermediate scenario might be a "counterforce strike," that is, an attack on military targets. A counterforce strike might lead to massive destruction in relatively unpopulated locations such as missile bases and airfields. The consequences for the public would still be disastrous, since the radioactivity released from such an attack would drift downwind to cities and because many military targets are located near cities.

Another attack strategy might be focused on critical elements of the economy. The goal in such an attack would be to damage the productive capabilities of the nation with relatively few casualties to civilians, and without explicitly attacking military bases. An example of such an attack would be one focused on oil wells. This kind of attack was examined in *The Effects of Nuclear War*.[6]

Finally, an attack seeking to produce maximum long-term impact with a few weapons might conceivably focus on nuclear reactors. Reactors contain enormous quantities of radioactive elements, many of which have very long half-lives. One result of such an attack would be that the time that would have to be spent in a shelter would be substantially more than that anticipated in most other scenarios.

This listing of possible modes of nuclear warfare is illustrative only. It serves to indicate that there are very many different situations which could result from the explosion of nuclear weapons over the U.S. For some of these situations modest civil defense programs could save many lives. For other situations only a massive civil defense effort could be effective. And for other scenarios it is hard to imagine any civil defense program that would be of much help. The extent of the destruction in some nuclear war scenarios is so vast that the resources of the U.S. would be inadequate to permit the nation to recover, regardless of the civil defense investments made.

Criteria

While the destructiveness of nuclear weapons is far beyond anything seen in any previous war, the general criteria for civil defense effectiveness are not significantly different from those in the past. I review some of the key features of the criteria discussed in a recent General Accounting Office report.[7] My discussion is general and is designed to serve as an introduction to the detailed analyses presented in subsequent chapters (see especially Chaps. 3 and 7).

> *Protection of civilians.* Defense measures can be thought of in terms of systems designed to cope with several categories of attack: (a) one or a few nuclear weapons arriving without warning; (b) warheads arriving after a period of increasing tensions (allowing time for individuals to move to shelters or to relocate); (c) continuing threats lasting weeks, months, or even years; (d) simultaneous attacks on many parts of the nations, making external aid unlikely; (e) large-scale attacks with long-term consequences such as nuclear winter. A comprehensive civil defense program should be designed to flexibly deal with all of these instances. In practice, the effects of a full-scale war may be so vast that effective measures would become impossible. There may be insufficient resources in the nation to permit building a system that would allow reconstruction of the society at all. Smaller-scale civil defense measures could yield some protection (e.g., against accidental missiles or terrorist attacks).
>
> Given warning, a shelter system could save large numbers of lives. Without protection a one megaton air burst is likely to kill most persons

within a radius of about eight kilometers. A shelter hardened to 10 psi (pounds per square inch) would provide protection even at ground zero against a one megaton bomb exploded at an altitude of 3600 meters. (This is the altitude which would be selected if the attacker's goal was to maximize the area experiencing an overpressure of 4 psi—a pressure high enough to cause extensive damage to unprotected persons, and to buildings not specially designed.) A shelter hardened to 100 psi could protect against a ground burst almost to the crater lip, at a distance of perhaps a half mile from ground zero for a one megaton bomb. An extremely hardened system could provide short-term protection for almost 99% of the people who would otherwise be exposed to the initial effects of a ground-burst nuclear explosion, and some advocates have argued that the U.S. should consider installing such systems (see Appendix B). Protection over the longer term would require many other elements in addition to hardening (e.g., stores of food, water, and medical supplies). The construction of hardened shelters is extremely controversial, since it would be expensive and disruptive, and would have major implications for the structure of the society and for our interactions with real or potential adversaries (see Chaps. 3, 6, and 7).

Protection of nonmilitary resources. A comprehensive civil defense program must be capable of restoring the agricultural and industrial capacity of a nation after a nuclear attack. It does little good to protect a population from the immediate effects of nuclear war if there is little prospect for long-term survival and recovery. Means must be provided for rebuilding a destroyed infrastructure. A limited attack might focus on particularly vulnerable elements of the economy such as utilities, oil refineries, transportation and communications networks, chemical and steel plants, etc. If the effects of nuclear winter are to be taken into account, then fuel, food supplies, seeds, etc., must be stockpiled in sufficient quantity and with sufficient protection to permit holding out until long-term climatological effects have moderated.

Communicate a sense of resolve. A functioning civil defense plan sends signals to an enemy. One argument sometimes used against civil defense is that such a program may indicate that a nation is preparing to attack or is prepared to accept an attack. These considerations are particularly clearly seen in discussions of the Soviet civil defense measures and their nuclear policy. A civil defense plan involving crisis relocation might either increase or decrease the likelihood of use of nuclear weapons. Such effects need to be balanced against the protection that civil defense measures can offer. Communication of national resolve might also serve to defuse a conflict situation, particularly one involving limited threats by nuclear terrorists.

Postattack recovery. An effective civil defense program must deal with the entire spectrum of problems that will face the nation following a nuclear war. Food and medical supplies must be provided, agriculture and industry must be reestablished. The postattack environment will be extremely difficult even if only a few nuclear weapons are exploded over the U.S. With a massive attack the problems might prove insurmountable, particularly should nuclear winter occur. There has been virtually no national planning dealing with periods lasting beyond a few months at most.

Communicate with our own populace. The implementation of a massive civil defense program would to some degree place the nation on a wartime footing. Such actions taken in peacetime would change national attitudes and might thereby contribute to the likelihood of a conflict against which the technical measures are designed to protect.

Non-nuclear war considerations. The previous categories focus on civil defense against *military* threats. Any civil defense program with these capabilities will necessarily also have capabilities to provide protection against nonmilitary natural disasters, such as nuclear reactor accidents, hurricanes, toxic releases, earthquakes, etc. One must be careful, however, in evaluating the synergism between civil defense measures designed for these different purposes. In general, many civil defense measures designed to provide protection against nuclear war will be useful for other types of disaster, but measures that are not specifically developed with nuclear weapons in mind will be of only minimal relevance to nuclear civil defense.

Recent developments

In the past several years there have been several developments which make reexamination of the role of civil defense in our national defense posture appropriate. These are (a) rejection of the Crisis Relocation Plan, (b) nuclear winter, and (c) the Strategic Defense Initiative (SDI).

Crisis relocation plan

Since the Carter administration, Crisis Relocation Planning (CRP) has been the cornerstone of U.S. civil defense planning (see Chap. 7). CRP assumes that nuclear war would not appear "out of the blue," but that there would be a period during which international tensions increase. This would provide time during which citizens could be evacuated from cities to less populated areas and critical resources could be protected.

The problems of moving large numbers of citizens over substantial distances (tens to hundreds of kilometers) in a few days and providing for their needs are clearly considerable. Further, there would be a great amount of dislocation associated with such a move in terms of financial loss, accidents, fires, theft, prospects for vandalism, etc. It is far from clear how long such relocation could be maintained. Further, an enemy could, to some degree, overcome such a system by selective retargeting or even by attacking during the process of relocation.

The Reagan administration decided that CRP is an inappropriate response to nuclear warfare considerations, and has moved to discontinue the program. The budget of the Federal Emergency Management Administration (FEMA) has been substantially curtailed. The history of U.S. civil defense activities is reviewed in Chap. 3, and CRP and the role of FEMA are discussed in Chap. 7.

Nuclear winter

Nuclear winter refers to long-term climatological changes which might result from the explosion of nuclear weapons over regions, primarily cities, with large quantities of combustible material. It has been estimated that a very small portion of the worldwide arsenal of nuclear weapons—perhaps as few as a few thousand weapons—exploded over large cities and forests, might lead to severe climatic change. The mechanism is injection of small particles of smoke soot and dust into the stratosphere. These particles would change the optical properties of the atmosphere so as to alter the Earth's energy budget, and to lead to cooling (or, in some cases possibly to heating) of the Earth's surface. In some studies, decreases in temperature of 20–40 °C for periods of several months to a year are forecast, with the effect being especially severe in the interior of continents. If smaller numbers of nuclear weapons were exploded, nuclear winter might not be triggered; if triggered, the effects would be less. Since the time for stratospheric particles to circle the globe is a few weeks, such effects would influence the entire northern hemisphere and might well extend to the southern hemisphere (see Chap. 9).

If a war were to take place at the beginning of the growing season (in the spring) then an entire year of crops might be lost. Such effects were actually observed—though at a level far below what a nuclear war might do—in 1815 when the Tambura volcano exploded in the South Pacific, leading to loss of crops throughout Europe and "the year without a summer" in New England. Nuclear winter would affect (at least) all northern hemisphere nations—combatants and noncombatants alike.

The recognition of the possibility of a nuclear winter situation developing places new constraints and requirements on a workable civil defense program. In Chap. 10 we review some of the ways in which historic civil defense planning would have to be modified to address nuclear winter concerns.

Defensive strategies: The promise of a shift away from offense-dominated strategies

President Reagan, in his "Star Wars" speech given on March 23, 1983, proposed that the nation undertake a program to shift our national military strategy from one of a balance of terror (Mutually Assured Destruction) to a defensive strategy. If such a move could be accomplished it would represent the first major change in nuclear politics in several decades. The technical challenges of ICBM defense are substantial, and it is important to remember that an effective defense would also have to deal with other attack systems including submarine-launched missiles, bombers, cruise missiles, etc. (areas not mentioned in President Reagan's speech). The nation is now in the process of evaluating possibilities and we can anticipate that many technically promising options will be identified. A defensive system which would destroy nuclear weapons entirely, or cause them to detonate outside of the atmosphere, would not only decrease damage to cities but would decrease the chance of nuclear winter occurring. If a partially effective defensive shield (possibly made feasible by arms control agreements) capable of defending cities, as well as military facilities, can be achieved, then the justification for civil defense measures may be greatly strengthened.

A world where we can "live and let live" (in the language of Dyson[4]) may be feasible (if it is feasible at all) only with a combination of innovative diplomacy, defensive technology, offensive arms limitations, and civil defense. This direction for development of nuclear strategy represents the first real prospect for change in many decades. It could lead to increases in nuclear stability, and reductions in overall risk. Or it could do the opposite. There can be no doubt, however, that regardless of how the nuclear arms race evolves, there will continue to arise important issues relating to civil defense.

References

1. James Turner Johnson, *Can Modern War Be Just?* (Yale Univ. Press, New Haven, 1984). See the references in this book for guidance to this extensive literature.
2. National Conference of Catholic Bishops, *The Challenge of Peace: God's Promise and Our Response*, (Office of Publishing and Publication Services, United States Catholic Conference, 1312 Massachussetts Avenue N.W., Washington, D.C., 1983).
3. James Reston, Jr., "A Reporter at Large: You Cannot Refine It," *The New Yorker Magazine*, January 28, 1985.
4. Freeman Dyson, *Weapons and Hope* (Harper and Row, New York, 1984).
5. Kurt Vonnegut, *Slaughterhouse Five, or the Children's Crusade* (Dell, New York, 1977).
6. Office of Technology Assessment, *The Effects of Nuclear War* (Allanheld, Osmun and Co., Totawa, N.J., 1980), issued by the Office of Technology Assessment, Report No. OTA-NS-89, 1979.
7. U.S. General Accounting Office, *The Federal Emergency Management Agency's Plan for Revitalizing U.S. Civil Defense: A Review of Three Major Plan Components*, Report to the Chairman, Subcommittee on HUD/Independent Agencies, Committee on Appropriations, U.S. Senate (U.S. GPO, Washington, D.C., 1984), Report No. GAO/NSIAD-84-11, 1984.

Chapter 2
History of American attitudes to civil defense

Spencer R. Weart

Civil defense is a combination of physical and social factors: the best-designed home fallout shelter is useless if nobody will build one. Any consideration of civil defense must therefore include some input from the social sciences. Our chief tools here are opinion polls, which show what people say they think about defense, and history, which shows what people have actually thought and done in the past. These approaches offer only the faintest predictive power, but they give the only evidence we have on some crucial questions.

The idea that warfare might destroy civilization, perhaps even exterminate humanity, was already familiar to the public in the 1930s. Atomic bombs themselves were occasionally predicted, but most people thought the idea either pure fantasy or a problem for a future too distant to merit serious thought. Rather, the public felt it would be chemical explosives and poison gas that might bring a new dark age. The image of a new world war with these conventional weapons, leading to utter destruction and barbarism, was widespread in popular journalism and stories, and even in a movie, *Things to Come*. During the 1930s this imagery may have retarded the acceptance of civil defense in the most threatened nation, Great Britain, where many argued that precautions would be futile. Others, especially on the Left, maintained that "Air Raid Precautions" would tend to provoke war, while still others pointed to the social difficulties of a program where only the wealthy could afford shelter. These arguments were met with vigorous counterargument in favor of civil defense, notably from the renowned Communist scientist J. B. S. Haldane. All the reservations were set aside after the 1938 Munich crisis, and every major European nation rushed to build up civil defense. The experience of the 1940 bombings of Britain, and intensely optimistic government wartime propaganda, convinced most Americans that civil defense was a sound undertaking. Ordinary citizens felt they could do what was needed to protect themselves. The failure of civil defense to protect German and Japanese civilians from the great 1944–45 air raids had little obvious impact on American thinking. This situation began to change only after the Hiroshima and Nagasaki bombings.[1]

First impressions, 1945–1953

Beginning in 1945 not only journalists and science-fiction writers but even the most sober leaders, from senators to popes, spoke of a possible end of civilization, if not indeed a doomsday—what President Truman and many others said might be exactly the Armageddon prophesied in the Bible. As sociologist Edward Shils remarked, atomic bombs made a bridge across which apocalyptic fantasies, marching from their ancient refuge among fringe groups, invaded all of society.[2]

Polls confirmed that Americans now felt personally threatened by uncanny forces. A majority said they believed that another world war was likely within their lifetime, with a real danger of their own families dying in atomic attacks. Nearly everyone thought that if another world war did come, most of the people in the world's cities might die. Nevertheless, most people said they were not urgently worried. For one thing, Americans at that time trusted in their authorities: the experts would surely do whatever was needed to control the dangers of atomic energy.[3]

Many responded to pollsters by saying that they just did not know. The whole affair seemed incomprehensible, a matter only scientists understood, something far removed from daily life. Besides, the bombs were still owned only by the United States, and one would not have to worry about an enemy attack except in some hazy future. Typical was the Texas cattle rancher who admitted that he was not safe against the dangers of an eventual war, but who remarked, "Most of my friends are more interested in this year's calf crop." The same people who admitted that they were in deadly peril declared with the next breath that they were not worried. Asked questions about foreign affairs, most people did not even bring up atomic bombs at all unless asked. These attitudes tended to linger even after the Soviet Union exploded its first atomic bomb. From the late 1940s on, close observers began to suggest that the public was indeed anxious, but concealed its fears. People seemed to look on atomic energy with such awe and horror that they gave up on it. Like death itself, atomic energy seemed to be so far beyond what a person could grasp that there was no point in wasting time worrying about it. A psychologist speculated that people were so afraid of the almost supernatural force of atomic bombs that "we hide our heads in the sand."[4]

Concerned about the seeming public apathy, a number of groups, pursuing various objectives, deliberately encouraged fear of nuclear war. The U.S. Air Force warned of vast devastation unless citizens joined the Ground Observer Corps to detect enemy bombers sneaking in; meanwhile "atomic scientists" and others tried to tell the public that without more international cooperation, dreadful explosions—which they described in horrifying detail—would annihilate every city. But the most important effort to bring worries of nuclear war into the open came when the American government launched a civil defense effort.

Talk about civil defense suddenly became widespread in newspapers and magazines when the Korean War broke out in mid-1950. What if the Soviet Union came into the war, and managed to drop its few atomic bombs on America? To deal with this problem President Truman created a Federal Civil Defense Administration (FCDA) which soon had a staff of over a thousand people. State and local civil defense agencies proliferated, with thousands more people coordinated by the FCDA. Enormous effort was spent arguing over and working out the boundaries between federal and local, and between civilian and military, responsibilities. Officials hoped to mobilize fifteen million volunteers to help out after the bombs dropped. The FCDA's motto would be, "Survive, Recover, and Win."[5]

The main job of the civil defense agencies was to train volunteers and other citizens in what to do when the bombers came. That meant telling as many people as possible about the havoc that atomic bombs could cause, and telling them in gruesome detail. This training effort brought American officials face to face with the problem of public fear of nuclear war.

In the beginning experts worried that the handful of bombs the Soviet Union owned in the early 1950s might bring less direct damage, if they were dropped, than might come from panic. Although true mob panic is in fact a rare phenomenon, in most Americans' minds there was a close connection between war and social breakdown. The idea caught on that atomic war would mean terrified mobs and cities disintegrating into chaos. By 1953 the American press was using the word "panic" fourteen times more often than in 1948. The horror movies that became popular in the 1950s, featuring giant ants or dinosaurs brought forth by nuclear explosions, almost always had scenes of stampeding mobs, sometimes seeming like more of a threat than the monsters themselves. To ward off such mass fear, experts recommended emotional "inoculation." By gradually exposing the public to images of nuclear war, civil defense experts hoped that people would get accustomed to horrid sights and not run around shrieking when the bombs came. Along with training citizens, then, inoculation was a second reason to broadcast the ugly facts about what atomic bombs could do.[6]

Whether or not they admitted it, American civil defense officials had a third reason to spread frightening images. Congressmen often said that civil defense was an important part of national security, but they never voted sums of money to match the enormous task. Through the 1950s the FCDA's budget appropriation was typically about one-fifth of the amount that Presidents requested. To make things harder, not nearly enough volunteers stepped forward. Civil defense leaders constantly worried over this "criminally stupid" lack of support. If only the public could be made to really see the dangers, would they not be more supportive?[7]

During the 1950s civil defense officials, seeking to promote training, emotional inoculation, and public support, spread images of nuclear disaster more efficiently than even the atomic scientists had done. In 1951, the first year of its work, the FCDA handed out twenty million copies of a pamphlet, *Survival Under Atomic Attack*. Meanwhile its movie of the same name sold more prints than any movie in the history of the industry. Traveling "Alert America" exhibits, with vivid pictures of atomic bomb destruction, reached over a million people. Other films, slide shows, and lectures came into every community in the United States. The armed forces vigorously aided this activity, for example inviting reporters to come aboard bombers in mock attacks on American cities.[8]

From the White House itself, one of President Eisenhower's aides, James

M. Lambie, Jr., promoted civil defense. He was aided by the Advertising Council, a private group set up to aid worthy national causes such as blood donation drives, Smokey the Bear, and the Cold War. The Council handled civil defense campaigns throughout the 1950s, soliciting free space in newspapers, in bus and subway advertising placards, in radio station breaks and so forth, and filling the space with messages crafted by Madison Avenue experts. Millions of dollars worth of advertising was donated, costing the government only a few thousand dollars a year for out-of-pocket expenses. Lambie, in common with many other Eisenhower supporters, particularly wanted to use this apparatus to spur Americans to greater efforts in the Cold War. He worried that unless he could make the public face the terrible danger of Communism, they would demand "dangerous reductions in our expenditures for armaments."[9]

At this time government propagandists could count on whole-hearted cooperation from the press and other groups. Besides the free advertising space, American newspapers printed tens of thousands of civil defense items in the early 1950s, and every major magazine published articles. Every radio listener from 1953 on got used to hearing not only civil defense advertisements and news but also tests of Conelrad, the emergency network. Towns set up air-raid warning sirens and tested them at intervals. It seemed that everyone wanted to help. For example, the National Automobile Dealers Association sponsored *Escape Route*, a short film about the value of a family car for escape or shelter under nuclear attack.

Eisenhower's war games

Most impressive of all were a series of "Operation Alert" exercises that involved the entire nation. In a 1954 drill, citizens of over fifty cities in the United States and Canada obediently sought shelter when sirens howled the warning, leaving streets eerily deserted. Larger drills followed in subsequent years. "125,000 KNOWN DEAD, DOWNTOWN IN RUINS" screamed the front page of the 20 July 1956 issue of the *Buffalo Evening News*—one of many civil defense exercise specials that newspapers printed to announce nationwide catastrophe, complete with chilling pictures of demolition. Even without that, newspaper photographs of streets deserted during the drills offered a somber vision of a world without people.

Operation Alert was a major job for the Eisenhower administration. The Cabinet spent long hours discussing the details of each exercise—the evacuation of Washington staff to secret command centers, the issuing of mock "emergency proclamations" on everything from medical supplies to banking, and other practical matters. To Eisenhower it was simple prudence to hold what he called "war games." Just as a peacetime army kept itself fit, so now, the General believed, the nation as a whole was on the front line of defense and must be prepared for combat.[10]

On only one occasion did the Cabinet seriously consider the chance that their war games might increase public anxiety. In a 1956 meeting, Secretary of Defense Charles Wilson suggested that the upcoming Operation Alert would "scare a lot of people without purpose." Wilson had been called by

Eisenhower to hold down the military budget; the President, a fiscal conservative, believed that in the decades of grueling struggle with Communism that he foresaw, too much spending on the armed forces would defeat itself by crippling the nation's economy. Now Wilson warned that the drills might ultimately inflate the military budget. For they would "strengthen and confirm the views of what might be called the fear lobby here in America—the people who might be said to have a vested interest in massive preparations for war." But Wilson and a few others who agreed with him were overruled.[11]

At the center of all the civil defense activity sat Val Peterson, head of the FCDA since 1952. He was a resolute man who had made his way up from schoolteacher to political science professor, newspaper publisher, and finally Governor of Nebraska, and he knew well how to speak to the American heartland. Peterson was dead earnest about his new job, working late into the night, worrying painfully over how to protect his country. He was emphatic about telling people the truth. For example, in a television talk that was also printed in *Collier's* magazine, he described just what an atomic bomb might do. "The heart of your community is a smoke-filled desolation rimmed by fires... Trapped in the ruins are the dead and the wounded—people you know, people close to you."[12]

Even more impressive than such writings was an effort at civil defense education held in the Nevada desert. In 1953 the government allowed hundreds of reporters to witness an atomic bomb test which included a civil defense exercise; three-quarters of the nation heard or read about it or saw it on television. Crouching in a trench close to the explosion was a network pool reporter, young Chet Huntley. Amid the confusion after the blast he described its overpowering shock and roar, exclaiming that the bomb was "the most tremendous thing I have ever experienced." According to a survey, millions of viewers also came away impressed by the bomb's destructive power.[13]

Of special interest were houses built nearby, "ordinary American houses" according to the networks' anchorman, Walter Cronkite. Fully dressed mannikins sat in the living rooms and kitchens. The idea of building houses had been promoted by the novelist Philip Wylie, a friend of Val Peterson's and a passionate advocate of civil defense, who had insisted that a grisly demonstration was needed to shake up people, to rub their noses in the problem of nuclear war. It did that. After the test, reporters took macabre photographs of mannikins twisted and broken amid shards of glass and collapsed lumber. Most Americans would have agreed with the name reporters gave the houses: Doom Town. For a repeat of the exercise in 1955, the FCDA built an entire village and christened it Survival Town. Reporters called it Doom Town anyway. While earnest civil defense people told television cameras that Americans did not need to be afraid, the reporters sounded less sure, and the visual images were unmistakable. A little boy mannikin torn to pieces, a typical house "blasted to smithereens" as a reporter said—those were not reassuring.[14]

Another civil defense program reached even more citizens. Although the United States government had little authority over schoolteachers, public anxiety ran so high in the 1950s that most schools gave at least token attention to civil defense. In the early 1980s, when I asked audiences of educated

American adults if they remembered what to do if a teacher sounded the alarm, about a third would put their hands up. The procedure recommended by the FCDA for a sudden attack was "Duck and Cover," and in tens of thousands of schools during the 1950s the children practiced ducking under their desks and covering their necks and faces. Some Catholic schools added a third step, instructing the cowering children to pray.

Many schools took more elaborate steps to prepare for the possibility that they would have some advance warning. Students practiced filing into a basement or evacuating to their homes. Typical of the programs was the one in Detroit, where the schools sprung various types of surprise drills and used the FCDA's *Survival Under Atomic Attack* as a fourth-grade text. Since the children might be separated from their homes, Detroit parents were asked to put names on clothing with indelible ink, and about half complied.

To be sure, identification by marking clothes was frowned on by some experts, who noted that "clothing can be destroyed by blast and fire." Even tatoos on the child's skin, one authority remarked, would not hold up against severe burns. Some of the nation's major cities therefore handed out heat-resistant dog tags to hundreds of thousands of schoolchildren. Occasionally Americans kept their tags to adulthood, tucked away in a drawer, all but forgotten.[15]

The training materials shown to these children, and for that matter to the several million adults who took civil defense courses, included strong images. The overall tone was deliberately confident, with calm actors and optimistic endings, but in an undertone the materials whispered a different message. For example, the pamphlet *Survival Under Atomic Attack* noted cheerfully that "lingering radioactivity... is no more to be feared than typhoid fever or other diseases that sometimes follow major disasters." The film of the same name featured a vivid scene of a window bursting and plaster raining down on a table set for a family dinner. Another short film pictured a surprise attack: a boy was lit by a sudden flash, hurled himself to the ground until the blast wave roared past, then got up and ran for shelter in a neighborhood suddenly converted to a landscape of rubble. No fictional television show or movie of the 1950s showed scenes of nuclear attack as shockingly realistic as that. Children who listened closely began to understand that even if they obeyed all the instructions, their chances of surviving an attack might not be very good.[16]

Not until long after, in the late 1970s, did some people begin painfully to talk about their childhood thoughts. A woman remembered that as a fourth grader in a small town she was sometimes frightened, particularly at night, if she heard an airplane drone overhead. She would sit up and beg, "Please don't let them drop the bomb on me!" One night in the 1950s, a young boy heard a siren and believed that the moment had come—"the most terrifying thing I can remember." College students recalled how, as children, they had harrowing nightmares of running for shelter from bombs.[17]

While children were presumably most susceptible to these anxieties, did not adults have similar feelings in the 1950s? One city official wrote Eisenhower to complain that civil defense exercises would "cause more heart trouble than this country has ever had," since "older folks now become frightened every time they hear the siren on an ambulance... They fear it is a raid." It is hard to know whether the official had put his finger on a general problem. Of course Americans showed a wide spectrum of reactions to civil defense as to everything else, reactions ranging from foolish confidence to blind despair. But what did the mainstream of the citizenry make of civil defense, and more generally of the whole idea of nuclear war?[18]

Back in 1951, an expert had warned that propaganda to inoculate the public against fear might cause the opposite of what was wanted, "emotional sensitization rather than adaptation," or even a sense of futility. Whether because of the FCDA's vigorous efforts, or because of all the other propaganda from other groups encouraging the same fears, heightened anxiety and a sense of futility did spread. Val Peterson himself came to see that the more he told the public, the less they wanted to hear. "Acceptance of civil defense," he wrote privately, means "acceptance of the reality of a threat so tremendous, so horrible, that there has seemed no defense."[19]

Images of nuclear war

When pollsters asked questions about what the United States might look like after a nuclear attack, most people could scarcely come up with an image. Everything after the bombs fell seemed like a blank. Many talked about panic, mass insanity, and blasted landscapes. A small minority came up with hopeful images of survival and recovery, while another minority spoke of the doom of civilization and even of all humanity. About a third gave purely emotional responses such as, "Oh my God it would be terrible, I can't imagine what it would be like, it would be horrible."[20]

One reason for the confused mixture of answers was an ambiguity in the question. If they tried to imagine a nuclear attack happening right then and there, most Americans' imagination failed them. But they still trusted that the problem was one for the future, and that America's defenses were strong enough for the present, which was fairly accurate through the mid-1950s. Most, too, might imagine themselves dying in a particular attack more easily than they could imagine their whole society disintegrating. They could accept their fate, believing that the larger civilization would carry on.

Surveys in 1954 found that nearly all Americans said that civil defense was a good idea for the United States as a whole; who could deny that it made sense to prepare to save as many lives as possible, just in case? Yet very few had themselves taken the least concrete action to protect themselves or their families in the event of war. Many people became curiously bland or evasive when asked to face the issues, and recruitment of volunteers fell far short of the goals. A typical comment came from a middle-aged Boston administrator: "If they expect me to fight a ten million ton TNT bomb with a shovel and hatchet and a bucket of sand, they're crazy."[21]

Except for the schools, most institutions did little but talk. The city that tried hardest, New York, got no further than to point out possible shelters with black and yellow signs—less a real protection against bombs than another reminder almost too scary to face. At a dinner party in Baltimore, Philip Wylie heard eminent citizens joke about the New York City shelter signs they had read about. When he told them that their own home city had the same signs on main streets and in department stores, the citizens were taken aback. Pursuing the matter, in a scratch survey Wylie found that only one of a hundred people he questioned in Baltimore could recall seeing the shelter signs that were all around them. "The atomic fears," he concluded, "were so great they could not let their minds recognize the signs. To do so would be to admit the reality of their fears: Baltimore could be erased from the earth."[22]

Val Peterson, soon after he took on his job as civil defense chief in 1952, gave up on any attempt to save the cities directly. His main reason was the coming of hydrogen bombs, which, he discovered, would irrevocably change the nature of warfare. By 1954 this realization reached the public as well. The only defense now would be to evacuate the cities before bombs fell. But many citizens, including some civil defense officials, doubted that there would be enough warning time to do that. Pulling no punches, Peterson openly admitted that life after a nuclear war would be "brutal, filthy, and miserable." Since he took up his job, he declared in 1956, he had been staring into hell. One night Peterson had a long talk with the Air Force general responsible for defending the United States against bomber attack. At the end of the evening the general hung an arm around the portly bureaucrat, saying, "Val, you and I have the two shittiest jobs in the American government and we will be the first two fellows court-martialed following the attack." Peterson agreed with all his heart. His labors had seemingly done less to mobilize Americans than to make them helpless with anxiety.[23]

The final effect of all the passionate talk about nuclear war was hard to define. In 1956, soon after the Cabinet argument over what Operation Alert might be doing to the public mind, Eisenhower asked for a "thoroughgoing study of the effect on human attitudes of nuclear weapons." Peterson called together a blue-ribbon panel of social science professors and other experts, who after some months delivered a top-secret report. The distinguished panelists confessed themselves baffled. They suspected that most people wanted to avoid nuclear war at almost any cost; beyond that they just could not say what the bombs would ultimately mean for people and for politics.[24]

In the late 1950s many things besides civil defense affected public imagery of nuclear war. The death of a crew member of the Japanese fishing boat *Lucky Dragon* after he was dusted with fallout from a hydrogen bomb test, and the subsequent debate over fallout, greatly increased anxieties about radioactivity. When Linus Pauling and others said that fallout from tests could slaughter millions, it was obvious that a war would be vastly worse— and not just for people in the cities but even for those hundreds of miles from a bomb explosion. The bestselling novel and widely seen movie *On the Beach*, joined by other fictional works and supposedly factual statements by respected figures, convinced many people that it was plausible that the whole world could be sterilized by nuclear war. Other movies featured giant creatures engendered by radioactivity (*Attack of the Crab Monsters*, leeches, scorpions, grasshoppers, etc.), which suggested that at best a postwar world would be a complete mess. A number of writers and film directors suggested that the world would revert to savagery: *I Was a Teenage Caveman*.[25]

There was something appealing in this. The idea of surviving, of becoming a new Adam and Eve, of shedding the restrictions of the old mass society, of walking freely into any abandoned house or going off into a green wilderness, of showing one's mettle in fights against monsters or wicked marauders, of helping to "build a new and better world on the ruins of the old" (as the back cover of a bestselling 1959 paperback put it), all this must have had fantasy appeal, for it sold a good many books and movie tickets. The muscular primitive with his scantily clad mate, prowling the ivy-clad ruins of Wall Street, was a fictional theme that could be traced back to the nineteenth century, but the fantasy now seemed a real possibility. This was an image at odds with the government's civil defense efforts—an image of self-centered individualism, altogether denying the survival of American democracy and society as a whole.[26]

It was fantasies, whether of a sterilized world, or a monster-infested world, or a world of a few brave individual survivors, that dominated the American fictional media. No attempts were made to portray the real consequences of nuclear war in a salable fashion. Popular nonfiction works on nuclear weapons were scarcely more helpful, for their total description of the probable postattack world was typically only a few general phrases and some statistics on how many might die.

Yet if most people would not look at realistic details of a world after the bombs, neither would most people consciously accept fantasies of doomsday nor of happy survivorship. Educated adults understood, if they thought about such things deliberately, that radioactive fallout would never really create instant giant mutants, that lone survivors would never really discover their ideal mates and found a new world, and so forth. Therefore two types of images of the world after a nuclear war competed for attention: the types with empty streets, lurching monsters, and cavedwellers, recognized as fantasy, and the vague and abstract notions of the truth.

Ideas about the reality were built upon the handful of blunt facts explained by atomic scientists and civil defense authorities, facts made meaningful by the tales of Japanese victims and by a few dozen photographs of Hiroshima and Nagasaki shown over and over. Probably many people were like the woman who reported that when the issue of war came up on a news program she might have a fleeting impression of the bombed cities. "What's imprinted on my brain is those photographs that I have seen, so if I ever did think of it I'd see a city destroyed and blackened and burned." By the mid-1950s many recognized that every city could look like that: the whole world as one enormous Hiroshima. Another report to the government by a panel of leading social scientists, this one quietly assembled by the National Academy of Sciences in 1962, said that many citizens expected "physical destruction to be almost universal and the post-attack world to be a hopeless shambles, in which everything worth living for will be irretrievably lost."[27]

Polls gave more details. Taken together, the results could be described in terms of how eight representative Americans felt in the early 1960s. One of the eight was certain that nuclear war would mean the end of all life on earth. At the other extreme, one was confident that the United States could come through without much damage (further surveys showed that this segment of the public knew little about nuclear weapons). A third representative citizen expected the end of civilization, or at any rate unbearably brutal conditions for the survivors. The other five, in the middle, felt that the United States had a chance of eventually rebuilding its society, but they expected great destruction first. Four of the five, and these were the half of the public who were generally best informed, thought that their own towns would be annihilated, or if not, they expected to be personally harmed by fallout from more distant bombs; overall they felt their own chances for survival were "poor" or "so-so."[28]

The 1962 social science panel noted that most people were ignorant or confused about what might happen. As the woman who spoke of the Hiroshima photographs explained, her image was "not elaborated." Up to a point, most people could understand what nuclear war might really mean, but that point was reached after a few sentences or brief impressions. Beyond that the imagination rebelled. Because the image of realistic destruction was rarely looked at squarely, it did not become vivid enough to entirely displace the image that spread across the background; from the corners of their eyes the public continued to glimpse lurid visions of an empty planet, radioactive monstrosities, or a new Adam and Eve with backpacks. None of this could lead to realistic thinking or effort towards civil defense.

The fallout shelter crisis, 1960–1962

The Kennedy administration refused to accept the public apathy regarding civil defense. When the new men took office, they understood that ballistic missiles were coming: within the next few years, the Soviet Union would for the first time get the capability to inflict great harm on the United States. The thought became urgent in 1961 when Khrushchev threatened war over Berlin and Kennedy made an equally belligerent reply. To be sure, there had been previous nuclear war scares. The first was in 1950 when Truman suggested he was considering using atomic bombs in Korea, and others followed when one or another world leader hinted at the use of nuclear bombs in the fighting over Viet Nam, the Middle East, or the Formosa Straits. On each occasion the American public had reacted with extreme nervousness. But Khrushchev's Berlin ultimatum brought the worst nuclear scare so far, at the same time forcing Kennedy's team to think seriously about real war.

Already in 1957 the top-level Gaither Report to Eisenhower had laid out the terms for winning a nuclear war. A true victory could be possible only if the war could be limited mostly to counterforce exchanges—missiles against missile bases and airports. But even under these restrictions, far too many civilians would die indirectly, from fallout, unless there was an enormous program to build shelters. Edward Teller and others explained to the public that a shelter program might ultimately mean digging deep caverns under every city. Shelters from nuclear blast and fire, an entrance within a fifteen-minute walk for every citizen in areas at risk, would be very costly, but some felt they would be worth the price.[29]

Meanwhile cheaper steps could be taken. Ever since American civil defense experts absorbed the lessons of the first hydrogen bomb tests, and still more after the Gaither Report, they had campaigned for fallout shelters. Their pleas had been ignored. Since the mid-1950s the civil defense program had been in disarray, starved by local authorities and Congress, widely regarded as a nest of incompetents, scorned or ignored by citizens. When Teller insisted that the United States should spend tens of billions of dollars on massive shelters, Philip Wylie, among others, replied that by the time the caverns were finished, the enemy would have built weapons strong enough to blast them apart. The sometime missionary for civil defense admitted that the hydrogen bomb had made him switch sides. "Where once I felt national apathy was dangerous," he wrote, "I now feel it would be common sense."[30]

Civil defense officials, starved for public attention and respect, turned hopefully to the new Kennedy administration. As far back as 1949, the young Congressman Kennedy had argued for more civil defense, and he had repeated the argument during his 1960 Presidential campaign. Furthermore, Kennedy had promised to awaken the nation from apathy and to face down the Soviets—just what civil defense advocates had been wanting for years. They bombarded the new president with memoranda saying that a shelter program "could do much to awaken the country" to defense needs, and that shelters "would show the world that the U.S. means business." The memoranda gave pragmatic arguments too. By 1961 most Russian adults had received several days of instruction and the Soviet government had built many shelters. Whether or not that work was of any real use, the Soviet leaders seemed to believe that it was. At some future point they might become convinced that they could defeat an unsheltered United States in war or humiliate it in negotiations. According to this view, shelters were an essential part of any deterrent. But perhaps the most effective argument was the demand for shelters as "insurance." If war ever did happen, shelters would save tens of millions of lives. How could anyone refuse a chance to save lives?[31]

On top of all this came political pressure on Kennedy, much of it from Nelson Rockefeller. In 1957 Rockefeller had worked closely with Henry Kissinger on Rockefeller Fund studies of nuclear strategy which endorsed shelters, and meanwhile his concern was sharpened by advice from Teller. As governor of New York State, in 1960 Rockefeller proposed the most ambitious civil defense program ever put forward by a powerful American leader. If the legislature had swallowed it, the program would have made New Yorkers spend a billion dollars, requiring by law that every owner of a building would make an individual shelter while the State would pay for community shelters in schools.

While New York State was Rockefeller's biggest sounding board, he had another one right next to the President's ear as a leader of the United States Governors' Committee on Civil Defense. Meeting with Kennedy in early May 1961, the Committee demanded presidential leadership for a revived fallout shelter program. According to a secret summary of the meeting, Rockefeller explained that the program was needed "to stiffen public willingness to support U.S. use of nuclear weapons if necessary." In June the United States Conference of Mayors passed a resolution demanding better Federal civil defense leadership; a number of Democratic and Republican Congressmen agreed; and the chairman of the Joint Committee on Atomic Energy told Kennedy he was going to hold hearings to stir up interest in fallout shelters. Here was a political challenge that Kennedy could hardly overlook.[32]

Already in April 1961, when he had the nation's radio and television airwaves all to himself during a civil defense exercise, Kennedy took the chance to urge individuals to "plan to protect their own families." The campaign came to a climax on July 25 when Kennedy delivered a tense speech over national radio and television. The Berlin crisis, he hinted in dire language, was leading the world to the brink of war. A draft of the speech prepared the day before had offered Americans detailed advice on how to take shelter from fallout ("You will need water more than food," etc.). This sounded too frightening, and in the final version Kennedy only said he would ask Congress for funds to stock shelters with food, water, and first-aid kits. The President also announced that in the coming months he would "let every citizen know what steps he can take without delay to protect his family in case of attack."[33]

The speech shocked Americans like the sight of a ghost. The federal civil defense agency had already been getting increasing numbers of letters from citizens who wanted information on protecting themselves, but after the Berlin speech, mail descended in an avalanche. By early August, just before the Soviets began to build a wall between East and West Berlin, the agency was getting over 6000 letters a day, more than they had received in the entire month of January. Civil defense was discussed in Rotary Club and P.T.A. meetings, in school boards and churches, and for two months newspapers got more letters to the editor on shelters than on any other issue. The strong public interest stimulated the press, or perhaps the press stimulated the public; newspapers, magazines, radio and television carried roughly ten times as many items about civil defense as they had carried a year earlier, more than at any time since the Korean War. Typical of the press reaction was a summary by John Chancellor on NBC's popular *Today* television show: he said he was afraid of attack, but "ready to support a civil defense program that teaches me how to survive."[34]

The administration was slow to take up that challenge. They had planned on sending a booklet to every household in the nation to keep Kennedy's promise that he would offer advice on protection; an aide remarked that the booklet would be more widely distributed than any piece of literature since the Bible. But as the uproar mounted into the fall of 1961, Kennedy's men were taken aback by what they had started. They kept revising and delaying the publication for fear of setting off total panic.

Others jumped to fill the information vacuum. Local civil defense officials finally found an avid audience, but they were soon outdone in zeal by enterprising businessmen. Swimming pool contractors suddenly became experts on shelter construction, and manufacturers of everything from biscuits to portable radios discovered that they were already making survival equipment. A reporter who checked the Chicago telephone book under "Air Raid Equipment & Supplies" found one firm listed in 1960, eight in 1961, and fifty in 1962. Major retailers and fly-by-night shysters, fighting for a share in the new market, took to radio and television and encouraged public anxiety about the dangers of nuclear war. Banks set out free civil defense leaflets in their lobbies and placed advertisements in local newspapers inviting people to get a loan in order to build a shelter. One bank, evidently not too worried that war was imminent, advertised, "You may borrow from $200 to $5,000 with up to 5 years to repay." No doubt all this was, as one draft of the President's civil defense booklet put it, "in keeping with the free enterprise way of meeting changing conditions in our lives."[35]

Nuclear disarmament groups, and their followers among academics, scientists, religious leaders, and so forth, vehemently attacked fallout shelters. In the first place, they said, the shelters were useless, for even if someone did live through the first few weeks, he would emerge into a world of radioactive rubble, not worth surviving for. In the second place, civil defense seemed actively dangerous. The protestors claimed that building shelters would give the public a false confidence in using bombs, would convince the Soviets that Americans were making ready to attack, and so forth. This was the period when intellectual studies of thermonuclear strategy and deterrence were at their peak; arguments for and against civil defense, and every other aspect of nuclear war, were so carefully elaborated that debaters over the subsequent decades have found surprisingly little to add.

The debate turned even more bitter as critics began to notice that fallout shelters might indeed preserve certain lives—of farmers far from the cities and of suburbanites who could afford protection. Tenement dwellers would have worse luck. John Kenneth Galbraith, in an acid letter to Kennedy commenting on a draft of the civil defense booklet, called it "a design for saving Republicans and sacrificing Democrats." Some critics went further, saying that a civil defense program would impose totalitarian controls on America. Major science fiction novels on the paperback racks described desperate, regimented societies burrowed entirely underground.[36]

It was obvious that no matter how much money was spent there would always be more people than there would be spaces in totally safe shelters. The press became fascinated with the dilemma of the shelter owner. Most people who owned shelters felt that they should give entry even to strangers, but more press attention went to those owners (about one in four) who said they would fight off any intruder, and to the few who defiantly announced that they had guns and would use them. Reporters provoked a moralistic debate, extracting peculiar statements from clergymen as to whether or not Christians had the duty to shoot someone who tried to break into their family shelter. More pragmatic thinkers recommended that people who were locked out should find the shelter's ventilation shaft and block it. A California entrepreneur offered fake air vents which could be planted in the yard to lead the neighbors astray.[37]

In a December 1961 memorandum to the President, his aide Arthur Schlesinger, Jr., reported that he had canvassed the nation and found that the shelter debate was becoming the chief domestic concern, verging on hysteria. The spirit of do-it-yourself which had inspired the home shelter program was slipping, he said, into a mood of every-man-for-himself.[38]

As the months passed and the Berlin crisis cooled, the debate eased off. When the Kennedy administration finally distributed its long-awaited booklet in January 1962 they did not send it to every household but set out millions of copies in post offices; by then they had beaten the text into a featureless mush. The administration turned its civil defense officials to preparing evacuation plans and communal shelters, which could be done quietly without alarming the public, and which might give Democrats a sporting chance. Manufacturers of home shelters went out of business by hundreds, and talk of civil defense began to settle back toward the moderate level of the late 1950s.[39]

Cuba and stasis, 1962–1978

Before the fallout shelter debate entirely died out there came a new shock as Soviet missiles in Cuba brought the world to the edge of war. In the series of scares that had come every two or three years since 1950 this would be the last. When the crisis broke out in October 1962, fear reached a higher peak than at any time before or since. The largest building supplies store in Washington sold out of sandbags on the day after Kennedy's television speech announcing a blockade of Cuba, while civil defense leaflets were snatched up at the Pentagon. In other parts of the country supermarket shelves were stripped bare by food hoarders, and housewives stashed flashlights and jars of water in their basements. Those who had spent the most time thinking about nuclear bombs were among the most worried. Undersecretary of State George Ball told his wife to try to turn her basement into a home fallout shelter; strategist Herman Kahn carried a transistor radio everywhere he went; some American peace activists fled to Australia; the atomic scientist Leo Szilard flew to Switzerland.

Once the crisis was over, many of the people who had been worried turned away their attention as swiftly as a child who lifts up a rock, sees something slimy underneath, and drops the rock back. A survey of one group of university students in October 1962 found that most had some fear, and one in five expressed high fear of nuclear war—but half a year later, only one in thirty expressed high fear. A more extensive poll of Americans found about one-fifth in mid-1960, as the Berlin crisis gathered, who said they had given some thought to building shelters; this rose to half of the sample just after the Berlin crisis; but after the Cuban crisis, only a quarter could even recall thinking about building shelters. Correspondingly, in newspapers and magazines the space given to issues such as civil defense fell back to the pre-1961 level.[40]

Over the next years interest fell much lower. For example, the *New York Times* published over 350 news items on civil defense in 1961, but only about 70 in 1963, 20 in 1966, and 4 in 1969. The *Readers' Guide to Periodical Literature*, in its list of magazine articles under civil defense classifications, showed a similar collapse of interest. A treaty banning open-air bomb tests and the coming of *détente* between the United States and the Soviet Union no doubt had much to do with this. There was an international collapse of interest in everything having to do with nuclear war.[41]

In his astute 1961 memorandum to Kennedy on civil defense, Schlesinger put his finger on the process. For years, he recalled, disarmament groups had looked for some way to make the problems of nuclear war seem real and immediate to people. The Berlin crisis and the fallout shelter debate had done that. "When people read about American and Russian tanks confronting each other on the Friedrichsstrasse," he noted, "they feel they can do something about it for themselves—they can decide whether or not to build a shelter." The Cuban crisis renewed the urgency of such decisions. By 1963 almost everyone had made up their minds one way or the other. Most had made up their minds to do nothing. Despite all the talk in the newspapers and television, only about one in eight Americans took any practical war precautions during the crises. Only about one in fifty had built even the crudest kind of fallout shelter. That was still over a million sheltered families, yet they were a tiny minority compared with everyone else. Even the RAND strategists and local civil defense officials who worked professionally on fallout shelters usually did not build one for their own families.

The panel of social scientists assembled for the government by the National Academy of Sciences in 1962 inspected the psychological consequences of the shelter debate. The hardest choice, the panel noticed, was faced by people who believed that they lived in target areas; these were in fact a majority of the population. The only realistic choice such people had, if they wanted to survive, was to uproot themselves and go live in the countryside. Only a tiny proportion of families did that. A survey found that during the Cuban crisis a third of Americans had discussed with their families what they should do if a war started, but only one-twentieth recalled even giving thought to fleeing from their homes. Since bomb shelters too were seldom undertaken, that left one last possible decision: to do nothing.

Once they had made the decision, the Academy's report said, citizens faced the problem of "cognitive dissonance." Simply stated, people often will hold beliefs that do not logically agree with the way they act. If something forces them to pay attention to the contradiction, to the dissonance, then they will want to adjust either how they act, or what they believe—unless they can turn away from thinking about the matter altogether. I do not know how many people settled their thinking in such a way, for the survey research done on behalf of civil defense was fragmentary and uncertain. So was public opinion itself. Even the small minority who decided to build shelters, when samples of them were hunted down by dedicated researchers, turned out to disagree with one another and to hold ambiguous and self-contradictory beliefs. Not even the decision to do everything possible always went along with a belief in heroic survival. Overall, however, in the early 1960s the pattern of opinion and action was compatible with the panel's theory.[42]

A few Americans made the decision to refuse outright to protect themselves. They firmly held the corresponding belief that, as one expert put it, "Fallout shelters are rat holes. I'd rather go out in a flash than be burned slowly to a crisp." These were the small minority who accepted images of the most hopeless doom; in a sense they might even find that comforting, for it justified their refusal to attempt self-protection.[43]

Most Americans, however, did not accept such entirely despairing images of the possible future. They agreed that enough work could give them some chance of survival, and approved of civil defense in the abstract. But they chose to leave shelter work up to the government, as if the problem were irrelevant to their own actions. At the same time, about half of all Americans told pollsters they did not think there was much chance that a nuclear war would happen any time soon: there was no urgent call to do anything to save themselves, so their inaction was justified.[44]

When the wording of the questions was changed, the results hinted that the comforting belief in safety was now held not only by a majority but almost universally. One poll found the usual result that a third of Americans thought nuclear war was likely, but when the questioners went on to ask more directly, "How do you think things are going to turn out?" Only 7% of Americans flatly predicted a disastrous future. Here, then, was the public's answer. The question came down to a decision between possibilities. What would the real future, the future that a person actively prepared for, look like? Would it reveal one of the fantastic and horrible nightmares of nuclear destruction, or would it reveal familiar scenes of normal everyday life? The answer varied from one individual to the next, yet all the evidence shows that most people decided they did not need to worry about nuclear war.[45]

Or that they would not, they would simply refuse to worry. The disparity in poll results, depending on just how aggressively the question was posed, hints at curious difficulties and inconsistencies of thought. Ever since the early 1950s roughly a third of the public had admitted they felt substantial fear of nuclear war. These were the people who, if asked what came to their minds if they heard a group of jet planes or a siren, said they thought of bombs. The numbers did not change markedly after the Cuban crisis was over. In fact, from the 1950s into the early 1980s, responses on polls to questions about how likely nuclear war seemed, and how devastating it would be, showed only relatively minor changes—a matter of ten percent this way or that. In the few years after 1962 when attention to nuclear war in various types of publications dropped to a quarter or even a tenth of its previous level, this was not because of some great change in the way people felt about the bombs.[46]

The change in press attention did reflect something that could be detected in opinion polls, when the questioner did not probe directly for feelings about nuclear war but simply asked what problems came to mind. In 1959, three of every five Americans gave "nuclear war" as the most urgent national problem; by 1965 fewer than one in five thought of that, and a few years later still the number became negligible. In short, about as many people as ever would admit fears if asked about them, but they would no longer bring it up spontaneously.[47]

Another wave of concern, 1978–1985

Interest in nuclear war began to revive in the late 1970s as *détente* faltered, as the number of long-range bombs available to the Soviet Union followed the American pattern with a tenfold increase because of the advent of multiple independent warheads, and as the United States set out to build a new generation of weapons. The leap in attention to nuclear war was not a reflection of any gross change in public attitudes. The most complete series of polls, taken by Jiri Nehnevajsa and others over decades, showed that public thinking had not changed much. Expectations that a nuclear war would come within the next five or ten years had subsided during the 1960s and rose again in the late

1970s, coming near but not reaching the Cold War peaks of the 1950s; however, these changes were relatively minor. For example, at the low point in 1972 about 50% of Americans thought there was an even chance that nuclear war was coming, and in 1978 this had climbed to about 60%.[48]

Perhaps the most insightful survey was one held in the United States in 1981. Out of every ten adults, only three chose, "I don't think a nuclear war is too likely so I don't worry about it," while two chose, "I frequently think and worry about the chances of a nuclear war." The rest, that is, half the population, agreed with, "While I am concerned about the chances of a nuclear war, I try to put it out of my mind."[49]

Putting it out of one's mind was most evident when anyone tried to talk about civil defense. Over the years since the Cuban crisis, every now and then one group or another had tried to promote a strong shelter program, and others had promptly argued against it. The public paid no attention. Polls had always showed that most ordinary Americans were in favor of civil defense, but only as something that they thought the government was already taking care of so far as anything could be done. Public apathy was faithfully reflected in Congress, which after 1963 refused to give more than token funds to civil defense or even to debate it.

Concern among government and other intellectuals began to revive as the world's arsenals became more capable of counterforce war-fighting, the type of war in which civil defense could make the most difference. Attention from the press revived in the late 1970s. According to Nehnevajsa, civil defense suddenly got more articles and television coverage in 1978 than in the whole preceding decade, and the interest continued into the 1980s. Nevertheless his surveys found that less than one American in twenty had recently discussed such matters with anyone. In 1980, when I telephoned the civil defense office in my county, I gave its head a surprise; he said he had not had a single request for information about fallout shelters in a year or more. His office, shrunken from a staff of twenty at the time of the Cuban missile crisis to only three, was mainly occupied with plans for evacuation in case the local nuclear reactor failed. In my area, as elsewhere in the nation, the food that had been stocked in shelters back in the early 1960s had rotted and been removed.[50]

The Federal Emergency Management Agency, as the civil defense organization was now renamed, was well aware of the problems. Having endured decades of administrative upheaval, budgetary starvation, and opprobrium (see Chap. 3), the agency had learned to keep a low profile. When it attempted in the late 1970s and early 1980s to take some concrete steps toward evacuation planning, organizing areas to receive relocated people, the results were a replay in miniature of the 1950s. Whenever news of the effort reached the public, it stirred up anxiety, outcries, and increasingly well-organized opposition. Muckraking journalists who repeated startling statements from civil defense advocates, and the crusading Physicians for Social Responsibility, and local groups in a number of communities, all turned any attempt to prepare people for war into an opportunity to warn them to have nothing to do with anything nuclear.[51]

By 1985 apathy had returned: in the *Readers' Guide* and other indicators, the amount of press attention to nuclear weapons was plunging from its 1982 high point even more rapidly than attention had fallen away in the mid 1960s. Some of the attention, to be sure, had simply been diverted into the hypothetical future weapons of the "Star Wars" Strategic Defense Initiative. But although this plan, like all strategies that pitted nuclear weapons against one another, logically would require an expanded civil defense program, that component was fading from public awareness along with all the other realities of existing nuclear weaponry.

Ideas about the aftermath of a possible nuclear war had changed little over the decades. The image of a thorough doomsday became scientifically plausible in the early 1980s with "nuclear winter" calculations. But already in the 1960s a few scientists and science fiction writers had warned of disastrous climate changes following a war, and in any case the end of the world in ice was one of humanity's most ancient and familiar images. By the 1980s most people had long since decided whether they would equate nuclear war with such universal doom. At the other extreme, the "survivalist" movement attracted a brief rush of press attention in the early 1980s, centered on adolescent fantasies of individuals gunfighting amid the ruins, as portrayed in pulp novels, a series of popular movies, and the movement's magazine; unlike most shelter builders of the 1950s, many "survivalists" openly turned their backs on their local communities and repudiated hopes of rebuilding American society as a whole. Meanwhile the great majority of citizens continued to think vaguely of a nuclear war as a sort of global Hiroshima. In all these images there was little place for real civil defense.[52]

A 1981 American poll asked a question similar to one that had been asked just twenty years earlier, and the comparison of the results is fascinating. In 1961, slightly over 8% of the adult population felt their chances of living through an "all-out nuclear war" were good; in 1981, some 9% felt they had a good chance of living through "a limited nuclear war in which the Soviet Union attacked some of our military bases." In 1961, those who felt their chances in a war were "just 50-50" amounted to 40%, compared with 43% for the 1981 war. In 1961, 43% felt their chances would be poor; in 1981, it was again exactly 43%. Rarely have two polls taken so far apart shown such comparable results. By the early 1980s, Americans felt about a "limited" war precisely as they had felt about an "all-out" war two decades before. And since most people thought that once the first bombs dropped, there would be no limits, their beliefs had in fact scarcely changed.[52]

This fact pointed to a curious blind spot that only a few thinkers seemed to recognize. Many politicians and military planners, many articles and television productions—notably *The Day After*, seen by millions in 1983—and even the widely noted 1979 study of nuclear war by the U.S. Congress' Office of Technology Assessment, implicitly assumed something like a 1960s war. That was possible in the 1980s only as a rigorously limited and controlled exchange, a mere few hundred bombs out of the tens of thousands now extant. Hardly any work of fact or fiction among the thousands I have seen, aside from a handful of little-known technical studies, could bear to face the historical fact that nearly all past wars went on much longer than was expected when they began. Survivors of a limited war could not set about recovering after the first few weeks, for once the spell of peace was broken, whatever organizations remained might husband their remaining weapons and continue to bomb one another sporadically for months, perhaps years or decades, in a desperate struggle for survival. This sharpened the fundamental dilemma of civil defense public relations, the dilemma that had existed ever since the advent of hydrogen bombs: any realistic preparations for war must focus attention upon an image of a future so dreadful that few citizens could bear to spend much time contemplating it.[53]

Comment

In studying what can be done today about civil defense, the United States in the 1980s is probably more closely comparable to the United States of a generation ago than it is to countries such as the Soviet Union, with its totalitarian social controls, or Switzerland and Sweden, with their traditions of social conformity and armed neutrality. The history of the American public's images of nuclear war, and of attempts to influence these images through civil defense education, does not encourage massive future civil defense efforts. There is every reason to believe that a strong initiative in civil defense would have the same results today as in the Eisenhower and Kennedy and early Reagan years: political divisions, heightened anxiety, some scattered pockets of useful activity, and a final reaction of apathy and despair. Except for brief periods of immediate crisis, the public will continue to be dominated by the normal human tendency to avoid facing terrible possibilities, accompanied by scarcely acknowledged fantasies that are still less in touch with reality.

Note: This chapter is drawn from a forthcoming book, to be published by Harvard University Press, which will include further documentation and information.

References

1. *Things to Come* (distributed by London Films; released 1936). See W. Warren Wagar, *Terminal Visions: The Literature of Last Things* (Indiana University Press, Bloomington, 1982); Royal D. Sloan, Ph.D. dissertation (University of Chicago, 1958). J. B. S. Haldane, *A.R.P.* (Gollancz, London, 1938).
2. Edward A. Shils, *The Torment of Secrecy: the Background and Consequences of American Security Policies* (Free Press, Glencoe, IL, 1956), p. 71.
3. Leonard S. Cottrell, Jr. and Sylvia Eberhart, *American Opinion on World Affairs in the Atomic Age* (1948; reprinted Greenwood, New York, 1969); Stephen B. Withey, *4th Survey of Public Knowledge and Attitudes Concerning Civil Defense* (Survey Research Center, University of Michigan, Ann Arbor, 1954). For an overview of many polls, see Jiri Nehnevajsa, "Civil Defense and Society," Report No. AD-44285, Dept. of Sociology, University of Pittsburgh, 1964 (available from Defense Technical Information Center, Alexandria, VA).
4. "Atomic Bomb—Operation Crossroads" (CBS radio, 28 May 1946), transcript No. S76:0502, p. 12; Rensus Lickert on "You and the Atom" (CBS Radio, 30-31 July 1946), tape No. R76:0223-0224. Both the transcript and the tape are in Museum of Broadcasting, New York City.

5. For this and the following, see Sloan, Ref. 1, and Thomas J. Kerr, Ph.D. dissertation, Syracuse University, 1969. A short survey is Allan M. Winkler, Bull. At. Sci. **40** (No. 6), 16 (1984). For another overview with further references, see Harry B. Yoshpe, *Our Missing Shield: The U.S. Civil Defense Program in Historical Perspective* (Federal Emergency Management Agency, Washington, DC, Contract No. DCPA 01-79-C-0294, 1981).
6. Val Peterson, *Collier's* 100 (21 August, 1953).
7. Ralph Lapp, Bull. At. Sci. **9** (No. 7), 237 (1953), p. 241.
8. United States Executive Office of the President, National Security Resources Board, Civil Defense Office, *Survival Under Atomic Attack* (U.S. GPO, Washington, DC, 1950). *Survival Under Atomic Attack* (distributed by Castle Films, released 1951).
9. James M. Lambie, Jr., papers, Dwight D. Eisenhower Library, Abilene, Kansas; Lambie to Sherman Adams, 9 July 1953, folder "Candor (1)," Box 12, White House Central Files, Confidential File, Eisenhower Library.
10. For example, Minutes of Cabinet Meeting, 10 June 1955, Box 5, Ann Whitman Cabinet Files, Eisenhower Library.
11. Minutes of Cabinet Meeting, 13 July 1956, Box 7, Ann Whitman Cabinet Files, Eisenhower Library.
12. Val Peterson, Ref. 11.
13. "Atomic Explosion: Yucca Flats, Nevada" (CBS-TV, 17 March 1953), tape No. T77:0328, Museum of Broadcasting. For poll, see Withey, Ref. 3, pp. 73–76.
14. Philip Wylie, "Agenda for a Bull Session," to Brien McMahon, 28 January 1952, folder "Wylie, Philip," Box 719, Records of the Joint Committee on Atomic Energy, National Archives, Washington, DC. "Atomic Bomb Blast" (CBS-TV, 5-6 May 1955), tapes No. T78:0333-34, Museum of Broadcasting.
15. See special issue of *School Life* **35**, Suppl. (Sept. 1953). William M. Lamers, Nat. Ed. Assoc. J. **41**, 99 (1952).
16. See Ref. 8. *Atomic Alert* (distributed by Encyclopedia Britannica Films, released 1951). A fine survey is the film *No Place to Hide* (distributed by Media Study-Public Broadcasting System, released 1981). See also Albert Furtwangler, Bull. At. Sci. **37** (No. 1), 44 (1981).
17. SANE Education Fund, "Shadows of the Nuclear Age: American Culture and the Bomb" (WGBH-FM broadcast and tape cassettes, 1980), cassette No. IV. Edwin S. Shneidman, *Deaths of Man* (Quadrangle, New York, 1973). I draw also on numerous personal connunications.
18. Eddie McCloskey to D.D. Eisenhower, 25 March 1954, folder "133B Civil Defense," Box 656, White House Central Files, Official File, Eisenhower Library.
19. Irving L. Janis, *Air War and Emotional Stress: Psychological Studies of Bombing and Civilian Defense* (McGraw-Hill, New York, 1951), pp. 221, 224. Peterson to Nelson Rockefeller, 1955?, folder "Civil Defense Correspondence," Box 19, Lambie Papers, Eisenhower Library.
20. Withey, Ref. 3, p. 72.
21. Withey, Ref. 3, p. 42.
22. Philip Wylie, Bull. At. Sci. **10** (No. 2), 37 (1954).
23. 1956, quoted in Sloan, Ref. 1, p. 227. My translation of "worst (blankest) jobs" in Val Peterson, "Basic Concepts in Civil Defense Survival Planning," pp. 32–35 in Military Industrial Conference, *National Survival in the Atomic Age. Proceedings... February 9–10, 1956*, copy in folder "Civil Defense 1951–1956," Records of the Joint Committee on Atomic Energy, National Archives.
24. A. J. Goodpaster, memorandum of a Conference (19 August 1956), 30 August 1956, folder "Aug. '56 Diary—Staff Memos," Box 17, Ann Whitman Diaries, Eisenhower Library. Frank Fremont-Smith *et al.*, "The Human Effects of Nuclear Weapons Development," Report to the President, 21 November 1956,

folder "AEC 1955–56 (1)," Box 4, Ann Whitman Administration Files, Eisenhower Library.
25. These movies sometimes appear on television, and nearly all may be seen or a summary read at the Library of Congress Film Division, Washington, DC.
26. Blurb on cover of Pat Frank, *Alas, Babylon* (Bantam, New York, 1959).
27. SANE Education Fund, Ref. 17, cassette No. VIII. National Academy of Sciences, National Research Council, "Emergency Planning and Behavioral Research" (National Academy of Sciences, Washington, DC, 1962), p. 5, copy in folder "ND2 Civil Defense 6/62-3/63," Box 596, Central Subject Files, John F. Kennedy Library, Boston, MA.
28. Withey, Ref. 3, p. 72. Nehnevajsa, Ref. 3.
29. For the Gaither Report, and general background and bibliography, see Fred Kaplan, *The Wizards of Armageddon* (Simon & Schuster, New York, 1983); Michael Mandelbaum, *The Nuclear Question: The United States and Nuclear Weapons, 1946-1976* (Cambridge University Press, Cambridge, 1979). Edward Teller, Bull. At. Sci. **13** (No. 5), 162 (1957).
30. Philip Wylie, Bull. At. Sci. **13** (No. 4), 146 (1957).
31. Folders "Civil Defense," Box 295, National Security Files; "Office of Emergency Planning," Box 85, President's Office Files, and "ND2 Civil Defense," Central Subject Files, Kennedy Library.
32. McGeorge Bundy, memorandum, folder "Civil Defense 4/61-6/61," Box 295, National Security Files, Kennedy Library.
33. Folder "Berlin Speech," Box 60, Theodore Sorenson Papers, Kennedy Library.
34. Arthur I. Waskow and Stanley L. Newman, *America in Hiding* (Ballantine, New York, 1962), p. 33 and *passim.*
35. Reference 34, pp. 101-124 and chap. 10; Thomas Kerr, Ref. 5, chap. 5; folders cited above, Ref. 31.
36. John Galbraith to J.F. Kennedy, folder "Civil Defense 10-11/61," Box 295, National Security Files, Kennedy Library.
37. For example, "Gun Thy Neighbor?," *Time* 58, 18 August 1961. On all this see Ref. 34.
38. Arthur Schlesinger, Jr., "Reflections on Civil Defense," folder "Civil Defense 12/61," Box 295, National Security Files, Kennedy Library.
39. *Fallout Protection: What to Know and Do About Nuclear Attack* (U.S. GPO, Washington, DC, 1962).
40. Mark Chesler and Richard Schmuck, Public Opinion Quarterly **28**, 467 (1964); G.N. Levine and J. Modell, *ibid.* **29**, 270 (1965). See also Helen Gaudet Erskine, *ibid.* **25**, 478 (1961).
41. My counts, from *New York Times* index and *Readers' Guide to Periodical Literature.* See also Rob Paarlberg, Foreign Policy **10**, 132 (1973).
42. Cognitive dissonance theory is reviewed in M. Fishbein and I. Ajzer, *Belief, Attitude, Intention and Behavior: an Introduction to Theory and Research* (Addison-Wesley, Reading, MA, 1975). National Academy of Sciences "Emergency" report, folder "ND2 Civil Defense 6/62-3/63," Box 596, Central Subject Files, Kennedy Library. For polls see Nehnevajsa, Ref. 3. Also F. Kenneth Berrier, Social Problems 11, 87 (1963).
43. See collection of comments in Nehnevajsa, Ref. 3, pp. 285–291.
44. See especially, Erskine, Ref. 40.
45. Among other sources see Jiri Nehnevajsa, *Issues of Civil Defense: Vintage 1978. Summary Results of the 1978 National Survey* (University of Pittsburgh Center for Social and Urban Research, Pittsburgh, 1979).
46. See Ref. 45.
47. Paul Boyer, Bull. At. Sci. **40** (No. 7), 14 (1984).

48. Nehnevajsa, Ref. 45.
49. Gallup poll in *Newsweek* 35, 5 October 1981.
50. Jiri Nehnevajsa, *1982: Some Public Views of Public Defense* (Center for Social and Urban Research, University of Pittsburgh, Pittsburgh, 1983).
51. Robert Scheer, *With Enough Shovels: Reagan, Bush and Nuclear War* (Random House, New York, 1982) was widely quoted.
52. Climate change: e.g., Barry Commoner, *Science and Survival* (Viking, New York, 1963), p. 122; Roger Zelazny's 1967 story "Damnation Alley," made into the movie *Damnation Alley* (distributed by 20th Century-Fox, released 1977). 1980s movies: especially the "Mad Max" series. Magazine: *Survival Weapons and Tactics* (1981–).
52. Jiri Nehnevajsa, Ref. 50. See also Gallup poll in *Newsweek* 35, 5 October 1981; Bernard M. Kramer, S. Michael Kalick, and Michael A. Milburn, J. Soc. Issues **39**, 7 (1983).
53. Congressional study published as Arthur M. Katz, *Life After Nuclear War: The Economic and Social Impacts of Nuclear Attacks on the United States* (Ballinger, Cambridge, MA, 1981). *The Day After* (ABC-TV, 20 November 1983). See Jack H. Nunn, Parameters **10** (No. 4), 36 (1980).

Chapter 3
FEMA: Programs, problems, and accomplishments

John Dowling

This chapter gives a brief history of civil defense since 1950, with a more extended look at the Federal Emergency Management Agency (FEMA), the government agency responsible for the United States civil defense program. FEMA's role and mandate is examined, and the important civil defense programs dealing with the threat of nuclear war are discussed. The problems that FEMA faces with these programs, as well as problems FEMA faces in general are treated.

Introduction

The Federal Emergency Management Agency (FEMA) is responsible for the civil defense program of the United States, including both nuclear and nonnuclear disaster preparedness. In this chapter we deal with what FEMA views as its role and mandate, a short history of civil defense and of FEMA, what current FEMA plans are, and what problems face FEMA. This chapter draws heavily on the following reports: *The Federal Emergency Management Agency's Plan for Revitalizing U.S. Civil Defense: A Review of Three Major Plan Components* by the General Accounting Office (hereafter referred to as GAO-1984)[1]; *Issue Brief on Civil Defense* by Reynolds of the Congressional Reference Division[2]; and in-house reports circulated by FEMA[3] and GAO.

FEMA's role and mandate

The GAO-1984 Report on FEMA states "The purposes of the U.S. Civil Defense Program are to (1) save American lives in the event of a nuclear attack, (2) contribute to the United States' ability to deter the Soviet Union from an attack on the United States, and (3) improve the ability of the states and localities to deal with emergencies that occur as the result of natural and technological hazards."

Although Congress controls the purse strings and hence sets the role and mandate, it is important to understand just what FEMA views its role to be.

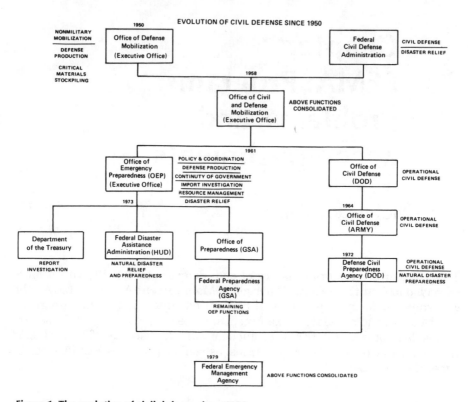

Figure 1. The evolution of civil defense since 1950.

An internal FEMA document gives the following as its "Mandate for Civil Defense":

> The basic purpose of government is to protect the lives of its citizens. Everything else springs from this immutable and fundamental responsibility.
> The policy of the Congress is to '...provide a system of civil defense for the protection of life and property in the United States from attack and from natural disasters.' (Federal CD Act, as amended.)
> The further policy of the Congress is that '...the responsibility for civil defense shall be vested jointly in the Federal Government and the several States and their political subdivisions.'
> The mandate of the Congress to FEMA is thus to work in cooperation with State and local governments to protect their people from disasters of all types.
> This protection must be provided where the people are—in local communities across the U.S.
> In sum, preservation of life is an inherent humanitarian responsibility of government to its people. The FEMA mission, as set forth by Congress in the CD Act, is to help save lives from emergencies of all types.

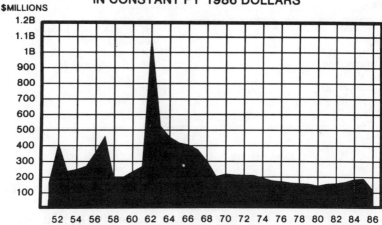

Figure 2. History of civil defense appropriations in constant fiscal year 1986 (FY86) dollars.

History of U.S. civil defense

A very brief history of civil defense is given by the two figures above. Figure 1 is from the GAO-1984 report mentioned previously and shows the evolution of civil defense since 1950. In 1979 FEMA was set up as an independent agency after an evolution from a multitude of agencies over the years. This was an evolution marked with more than its requisite share of the number of turf battles between government agencies and departments. A bewildering array of changes and bosses is a contributory reason to why civil defense has had so little clout and influence. The net effect has been to underfund and understaff the civil defense program with a resulting loss in effectiveness in carrying out what the agency believes to be its mission.

Figure 2 is from an internal FEMA document and details the budgetary history over the years. Some historic events important to civil defense are indicated in Table I. While there is not a strict correlation of events to appropriations, there are certainly very interesting parallels. Readers are advised to note that fiscal year appropriations are made in the year preceding the actual fiscal year.

Before we discuss FEMA today let us take a quick look at the history of civil defense since 1950. Those who wish more details should consult Blanchard's very readable history of American civil defense from 1945–1975.[4]

Truman administration

The Federal Civil Defense Agency was formed in 1949. The immediate stimulus was the detonation of the Soviet A-bomb, coupled with the remembrances of what happened in Hiroshima and Nagasaki. Congress held hearings on

Table I. Historical events pertinent to civil defense.

Date	Event
29 August 1949	First Soviet A-bomb detonated
24 June 1950	South Korea invaded
8 August 1953	First Soviet H-bomb detonated
1 March 1954	Bravo test: fallout on Bikinis and *Fortunate Dragon*
July 1955	Bomber gap
4 October 1957	Sputnik
Summer 1961	Berlin Crisis
October 1962	Cuban Crisis

civil defense which resulted in the Federal Civil Defense Act (FCDA) of 1950. The start of the Korean War in the summer of 1950 added impetus to civil defense planning. FCDA was the basic authorizing legislation for population protection programs. As a portent of budget problems to come, the first FCDA request for funds was cut from $403 to $31 million. The emphasis at that time was on in-place shelters as protection against very limited numbers of weapons. (This was before the days of an intercontinental Soviet bomber fleet.)

Eisenhower administration

Interest flagged and the Eisenhower administration did little with civil defense programs. But when U.S. and Soviet H-bomb detonations revealed the *previously unrecognized* danger of fallout and the increased destructive potential of thermonuclear weapons over fission weapons, there was a change in the approach to civil defense from in-place shelters to the evacuation of target areas. (There is an interesting parallel between this "unrecognized danger of fallout" to today's debate over the "unrecognized danger of nuclear winter.")

Kennedy administration

The period of international tension highlighted by the Berlin Crisis in 1961 and the Cuban Missile Crisis in 1962 created a peak in the Presidential, Congressional, and popular interest in civil defense. In 1961 President Kennedy asked for, and received from Congress, a supplementary appropriation of $208 million, in addition to the $87 million already appropriated for FY62 civil defense. In terms of 1986 dollars, the FY62 funding would amount to over $1 billion. This extra money was used for shelter surveys, and the marking and stocking of shelters with food and medical supplies. $695 million was requested in FY63 to continue this program, as well as to expand it to include incentives for building of shelters. However, no appropriation for shelter incentives was made, and civil defense funds for existing programs were cut to half of the FY62 levels. For FY64, congressional and popular enthusiasm had dropped even further and the request was cut from $346 to $112 million.

Johnson, Nixon, and Ford administrations

There was little interest in massive civil defense programs in the Johnson administration. Civil defense appropriations averaged $91 million yearly from 1965–1969. The Nixon and Ford administrations' interest in civil defense focused on "civil preparedness" to deal with natural disasters rather than on attack-related civil defense. The concept of "dual use"—using Federal attack-related resources in State and local natural disasters—was advocated as the best way to use available resources in the Nixon administration.

Carter administration

President Carter issued Presidential Directive No. 41 after administration studies came up with a role for civil defense in the strategic balance. This called for a gradually enhanced civil defense program that emphasized crisis relocation of population at particular risk from nuclear atta?. co presumably safer, largely rural areas (the "risk" and "host" areas), and continuity of government. This would enable us to respond, in kind, if the Soviets ordered an evacuation of their cities. The concept of dual use was again emphasized. These studies also resulted in President Carter's Reorganization Plan No. 3 of 1978 and Executive Orders 12127 (March 1979) and 12148 (July 1979), which created FEMA. The previously scattered Federal responsibilities to plan for, respond to, mitigate, and recover from emergencies of all types, natural, technological, or attack-related, were all consolidated within FEMA.

In 1980 Congress amended the Federal Civil Defense Act by adding Title V. This set forth Congress' legislative intent for an improved civil defense program emphasizing crisis relocation. Congress saw this improved program as enhancing the survivability of both the population and the leaders of the United States, reducing U.S. vulnerability, enhancing deterrence and stability, and reducing the chances of coercion by an enemy. Title V also called for attack-related and disaster-related resources to be used interchangeably.

Reagan administration

For the Reagan administration, as for the Carter administration before it, civil defense is an element in the strategic balance. Analysis of the role of civil defense in that balance led the Reagan administration to conclude that a moderately enhanced civil defense program could improve the overall U.S.–U.S.S.R. strategic balance.

The National Security Decision Directive (NSDD) No. 26 of 1982 stated the administration's concept of civil defense. NSDD No. 26 outlined the goals for the proposed seven-year $4.2 billion plan for an enhanced system, and enumerated the steps to be taken to implement those goals. This document echoed many of the ideas expressed by Congress in the 1980 and 1981 amendments of the FCDA. The latter amendment stressed that dual use was to be practiced in U.S. civil defense programs.

The unclassified version of NSDD No. 26 called for a "strong and balanced program of strategic forces including...strategic defense," and stated, "it is a matter of national priority that the U.S. have a Civil Defense program which provides for the survival of the U.S. population."

The Congressional Research *Issue Brief on Civil Defense* states the following:

> The proposed program was to:
> 1. Enhance deterrence and stability by 'maintaining perceptions that' the strategic balance 'is favorable to the U.S.'
> 2. 'Reduce the possibility that the U.S. could be coerced in time of crisis.'
> 3. 'Provide for survival of a substantial portion of the U.S. population in the event of nuclear attack preceded by strategic warning and for continuity of government, should deterrence and escalation control fail.'
> 4. 'Provide an improved ability to deal with natural disasters and other large-scale domestic emergencies.'"
>
> These policies were to be put into effect in three ways:
> 1. *Population protection*—crisis relocation of populations from high-risk areas to lower-risk areas with plans and deployment of supporting operational systems to be completed by 1989.
> 2. *Industrial protection*—analyses leading to funding decisions on a program to protect key defense and relocation support industries.
> 3. *Blast sheltering*—analyses leading to a funding decision on blast sheltering of the key workers in defense and relocation support industries.

The Integrated Emergency Management System approach (IEMS)

The IEMS approach was set up to deal with both attack-related and disaster-related emergencies. Previously, these emergencies were dealt with as separate problems with their own organizations and procedures. FEMA's reasoning was that both kinds of emergencies required similar responses so that capabilities for handling one set would be incorporated in handling the other.

For example, if there were to be a major evacuation of a city due to a serious radioactive release in a nuclear power plant, the evacuation plan developed for this emergency would be useful for evacuation in case of a nuclear attack. IEMS is eliminating the distinction between disaster- and attack-related activities. This process has met with some criticism. Cynics say IEMS "sells" better than does crisis relocation and they argue that the two types of emergencies are completely different and money, resources, and talent are being siphoned off to attack-related issues at the expense of disaster-related emergencies. Recent budget cuts at FEMA appear to negate these criticisms however.

Current civil defense programs

A partnership of the federal, state, and local governments is the basis for the National Civil Defense Program. Guidance, technical support, and financial assistance is provided by the federal government with full federal funding and assistance primarily for attack-related initiatives which include the following:

(1) *Radiological defense*: 52 full-time state-level Radiological Defense Officers (RDO), maintenance of 4.2 million radiological instruments purchased in the 1950s and 60s, and pilot production of newly developed instruments.
(2) *Shelter development and surveys*: surveys about 824 000 buildings to identify existing radiation shelters.
(3) *Nuclear civil protection planning*: developing plans to either relocate people or use best available in-place shelters.
(4) *Industrial protection*: developing cost data and test procedures for protecting vital industrial equipment in a crisis.
(5) *Essential worker protection*: identifies key industries and organizations that must operate and assesses costs to construct blast shelters for essential workers.
(6) *Miscellaneous programs*: these include shelter marking and stocking, producing ventilation kits, crisis shelter upgrading, and publishing emergency instructions to the public in local telephone directories.

The shared federal, state, and local funding and assistance is provided for dual-use initiatives such as the following:

(1) *Organization structure and emergency operations planning*: this supports 1300 state and 4800 local planners, as well as 1000 military reservists who provide emergency management assistance.
(2) *Warning and communications systems*: funds development, upkeep, and modernization of Emergency Operation Centers (EOC), development of survivable communications, and the broadcast station protection plans (BSPP).
(3) *Miscellaneous programs*: these include research on the technical and scientific base for IEMS, training and education, and the development of a National Emergency Management System.

The seven-year proposed civil defense plan

On October 2, 1981, after an administration review of civil defense programs and policies, President Reagan announced his intention to "devote greater resources to improving our civil defenses." This was part of his plan: "to revitalize our strategic forces and maintain America's ability to keep the peace well into the next century." In early 1982 Reagan signed a national security decision directive which stated that civil defense is an essential ingredient of U.S. nuclear deterrence and that it was a matter of national priority for the United States to have a civil defense program that provided for the survival of the U. S. population.

The civil defense objectives of this directive were almost the same as those called for by Title V. In order to implement these policies President Reagan gave the following directives:

By the end of 1989, complete the development of plans and the deployment of operational systems to provide for population protection, with priority being placed on population relocation during a crisis from U.S. metropolitan and other high risk areas to surrounding areas of lower risk.

Complete analyses and preparations required to make a funding decision on the protection of key defense and population relocation support industries.

Complete analyses and preparations to allow a funding decision on blast shelters for key industrial workers in defense and population relocation support industries.

FEMA then designed a revitalized civil defense program that was to be implemented between FY83 and FY89. This called for $4.2 billion to be allocated over seven years. This plan was designed to relocate the population from the risk areas during a crisis period which was expected to precede a nuclear attack and to provide the population with fallout protection and support in host areas—areas not likely to be hit directly in an attack. This seven-year plan was a composite of new civil defense activities and improvements to current FEMA programs. IEMS was another iteration on this development.

Problems with FEMA programs

We now look closely at a few of the major programs listed above. We also examine some of the problems with individual programs as well as with FEMA policy as a whole. Bear in mind that FEMA argues that many of the criticisms raised about their programs can be attributed to funding of FEMA at levels inconsistent with their mandate from Presidential Mandate No. 41.

Crisis relocation

Since 1978 crisis relocation has been the basis of the U.S. Nuclear Attack Civil Preparedness (NACP) program. The seven-year plan is also based on crisis relocation. This program has drawn much ridicule from opponents of civil defense and they feel it is laughable to try to evacuate major cities on the basis of a three- to seven-day warning period.[5,6] It looks like FEMA is quietly dropping the program for now. The official FEMA statement on this is taken from their booklet *The FEMA Budget in Brief FY* 1986:

Beginning in 1983, FEMA requested funds for the first year of a substantially enhanced Civil Defense Program, a program that would have included meaningful population protection in the event of attack on the United States. Because this element of the multi-year program was contested, FEMA's efforts over three years to obtain large increases for the program resulted in funding for only modest real growth. The program has been reduced in 1986 to a minimum level. During this time, we will continue to review the program and its elements. We, thus, regard 1986 as a maintenance year while we address the problems of public policy involved in the Civil Defense Program.

Table II. Status and types of crisis relocation plans (CRP's).

Areas	CRP's required	CRP's completed	Percent completed
Risk	625	243	39
Host	2,005	993	50
Combined	505	253	50
Total	3,135	1,489	47

To date FEMA has concentrated on getting states to develop crisis relocation plans to help people move out of the major high-risk areas in the event of a major crisis. Table II shows the status and types of crisis relocation plans (CRP's) completed by the end of FY83 (the situation has not changed much since then).

Part of the CRP program involves determining what shelters are available. Here is the latest status of the shelter survey (furnished by FEMA in a private communication). In FY84, the focus of the shelter survey shifted to multihazards instead of nuclear attack exclusively. The number of buildings surveyed in FY84 totaled 1 680 047. The buildings contained 181.5 million spaces that could provide public lodging during a crisis. Of the total buildings surveyed, 540 000 were found to contain 394.2 million spaces that could provide nuclear fallout protection. In FY85 the shelter survey continued to be multihazard oriented. Surveys for fallout protection spaces have only been conducted outside of risk areas. In FY86, surveys continue, but on a much reduced scale because of cuts in FEMA's personnel authorizations and budget.

After a CRP is completed there is supposed to be an enhanced plan which includes organizational relocation, work with essential industries, and stockpiling of essential food and medical supplies in host areas. But FEMA has been very slow in guiding enhanced planning and FEMA policy calls for the completion of all initial CRP's before enhanced planning occurs. Consequently, CRP's have become dated and useless. Another problem is that FEMA concentrated on CRP's for the smaller risk areas and left heavily populated areas for last. This policy initially directs civil defense efforts away from the most difficult cases—the areas where they have the greatest life-saving potential. The critical unknown (which is fast becoming known) is to what degree the state and local governments will participate in CRP. As of March 1985, seven states and some 120 localities representing some 90 million people had refused to participate in a CRP.[7] There appears to be poor coordination of CRP with other federal agencies. (Some CRP's might not be feasible because the Department of Defense plans may restrict interstate highways or other facility use. FEMA officials do not think this will be a problem, based on experiences in operations near various military bases. However, this author feels that a major evacuation will cause conflicts with such agencies as the Department of Defense.) CRP effectiveness is dependent on other factors also. Important ones are the industrial protection and essential worker shelter programs—programs which could greatly increase overall civil defense program cost. In general, crisis relocation seems to have ground to a halt because of public ridicule mentioned above, inadequate funding, and ineffective guidance and setting of priorities by FEMA itself. Finally, there appear to be serious legal and constitutional questions about CRP, and these difficulties have not been analyzed. For more detailed treatments of sheltering and of CRP see Chaps. 6 and 7, respectively.

Radiological protection

Radiological protection is essential to ensure postattack survival. It is designed to save the lives of the millions who would survive the direct effects of a nuclear attack, but who would otherwise succumb to overexposure to radiation. Radiological protection provides information, equipment, and technical advice essential to protect the population from exposure to radiation that could occur from nuclear attack. There are problems with the number of instrument sets, accuracy of equipment costs and cost production estimates, the status of current stores of equipment, and the adequacy of program staffing and implementation plans. Funding for radiological protection has been inadequate and falls far short of what is called for in the seven-year plan.

It is imperative that there be significant numbers and appropriately designed radiological instruments spread throughout each of the 3135 government jurisdictions. People in shelters have to be able to determine radiation levels. FEMA has attacked this problem in a variety of ways. They have made great strides in instrument improvement in developing two basic types of instrumentation: shelter sets and postattack recovery sets. Each shelter set contains two dosimeters (measures accumulated dosage received), one ratemeter (measures incoming exposure to radiation), and one charger (a small generator used to renew the dosimeter's electric charge). The postattack recovery set is the shelter set plus a more complex ratemeter.

A 1979 study by Oak Ridge National Laboratory recommended production of 10 million sets for population protection. FEMA reduced this number to seven million by eliminating sets for shelters in rural areas, and then reduced the number to 5.5 million due to budget constraints. FEMA hopes to reduce costs from $100 per set to between $30 and $40, but this is based on technological breakthroughs in instrument design, which may or may not occur (if radiological meters could go the way calculators did, the premise is not far fetched). Current estimates are also based on production in a federally owned facility in Rolla, South Dakota, using mostly less expensive labor from a nearby Indian reservation. In addition there are equipment inventory problems such as maintenance, vandalism, theft, and loss with the meters produced in the 1960s.

Radiological protection also involves an intensive and comprehensive radiological training series which was initiated in FY85. This new system responds to all radiological emergencies, involves a wide range of emergency services and facilities, and is broadening support and training facilities. This training series includes three different fundamental courses for radiological monitors, for radiological response teams, and for radiological officers. It also includes a radiological monitor instructor course. An integrated radiological emergency response system features a nationwide system of response personnel trained to deal with any radiological emergency in a very short time frame. This calls for 9600 radiological officers, 96 000 radiological response team members, and 1.2 million radiological monitors. These people are drawn from fire, police, public works, emergency medical and ambulance personnel, etc. The radiological officers serve as the primary point of contact. There have been staffing problems such as finding qualified people, finding slots at the various state and local governmental levels, and convincing local officials that long-term FEMA funding of these positions will continue. This program is based on a "cascade" training system concept: one level trains people, who then train others, and so on. There are problems here with training materials and unevenness in the quality and ability of instructors.

The GAO-1984 report had several recommendations on radiological protection. FEMA needed to better identify equipment levels and needs, review equipment distribution plans, update course materials, and develop a more accurate system for cost estimates. For a fuller discussion of radiation protection see Chap. 6.

Direction and control programs

Direction and control programs are needed to implement U.S. civil defense plans for crisis relocation and postattack recovery. In order to implement CRP it would be necessary to evacuate and relocate up to 145 million people. The government would warn people and tell them where to go, what to do, and how to protect themselves. After an attack officials would have to direct recovery activities and supervise the resumption of basic services and activities. Two basic programs to ensure this direction and control are Emergency Operation Centers (EOC) and the Broadcast Station Protection Plan (BSPP).

EOC's are protected facilities which would serve as local command posts. The program provides technical and operation guidance and up to 50% matching funds for upgrading and developing new state and local EOC's. At the end of FY83 only 350 of 3063 state and local EOC's met FEMA standards. In addition many local governments do not exercise direction and control plans, and there need to be increased levels of participation. FEMA has been very slow in developing plans for a national network of EOC's, that is, their number and location, and their guidance affecting EOC development is dated and needs improvement. Also in question are FEMA's cost estimates for upgrading and developing this network.

BSPP provides 100% federal funding for the acquisition and installation of facilities and equipment that protect selected broadcast stations against the effects of nuclear weapons. Some 2771 commercial radio and TV stations are to be selected from the more than 9000 stations participating in the emergency broadcast system. FEMA funds are used to develop a fallout-protected area within a broadcast station, to acquire adequate emergency equipment to operate in a radioactive fallout environment, and to provide protection against electromagnetic pulse (EMP). As of the start of FY85, 641 stations had been protected against fallout, but only 110 had EMP protection. But FEMA has not coordinated the station selection with broadcast coverage capability. Instead it has concentrated on getting a station in each jurisdiction. BSPP stations are expected to operate up to 14 days in a nuclear emergency. The GAO report doubted that the eleven stations they chose to visit could conduct effective operations after a nuclear attack. Again there has been a lack of a coherent, detailed set of procedures to explain what the program does, how it functions, and how it interacts with other civil defense programs. Both the EOC's and BSPP stations need to be inspected, and this is rarely done.

Conclusions

In his final chapter Blanchard argues that the problems of civil defense are not technical ones, but rather social and political ones. He discusses five major nontechnical determinants of U.S. civil defense policy:

(1) International crisis and change
(2) Quality of civil defense leadership and planning
(3) Congressional support and appropriations
(4) Presidential interest and support
(5) Defense policy

We briefly expand on the first four of these determinants; see Chap. 7 for a treatment of defense policy. The effect of international crises is obvious from Fig. 2, which has the dates of important international events entered onto the FY appropriations chart. The quality of the leadership has certainly had its effects. Almost everyone has heard of T.K. Jones's comment "It's the dirt that does it!" as mentioned by Scheer.[8] But an even funnier one is from Val Peterson, Eisenhower's director of civil defense in 1955, which Blanchard[4] (pp. 135–136) cites. "...it would be my plan to employ trenching machines and go along the public highways and dig miles of trenches 2 feet wide and 3 feet deep which can be dug at a cost of about 25 cents a running foot, and place people in those shelters." Then, "over the top of these trenches... I would suggest using boards and cover the boards with a foot or more of dirt." An alternative to the board covering, he continued, might be to place tar paper over the trenches so that "a person standing in one of those trenches could flap that thing every 20 or 30 minutes and shake that stuff (fallout) on the ground." Two citations from Blanchard[4] (pp. 218 and 450) show typical congressional level support. In 1958 Representative Johnson of South Carolina stated "one of the most glaring deficiencies of the Eisenhower administration is the way it has made civilian defense an abused stepchild of the Federal family. It would be hard to find anywhere a group of people with poorer morale than the employees of the Civilian Defense Administration, through no fault of their own." And Representative Montoya asked Director Davis (the civil defense head in 1974) about a conference held in San Antonio on crisis relocation: "Why did you pick on San Antonio, because of their expertise with the Battle of the Alamo?" In terms of appropriations Blanchard[4] (pp. 475) mentions that "...civil defense received roughly only 40 percent of its appropriation requests during the 25 appropriation years covered (1950–1975) in this study." In summary, Blanchard (pp. 492) states that "...high Executive (Presidential) interest (or lack of it) has been the primary determinant of civil defense policies and programs."

In this survey of some of the more important FEMA programs we find FEMA to be in the unenviable position of an inexperienced father who has to change a messy diaper for the first time. It is not a job he wants to do, but it has to be done, and done quickly and correctly. It is a messy job, one with unforeseen problems, and not very rewarding; but it is a job he is required to do with the best resources and talents he can muster. He eventually muddles through, and baby and father survive. FEMA has been given the job of saving lives in the event of a nuclear attack. FEMA cannot argue whether this is the appropriate problem to be working on, it just has to do it. And it has not been given the talent, the resources, or the commitment to carry out its mission. Within budgetary and staffing limitations it appears to be doing the best it can to solve a difficult problem, and, as we mentioned before, a problem that a number of people think is the wrong problem on which to focus.

References

1. U.S. GAO Reports: *The Emergency Management Assistance Program Should Contribute More Directly to National Civil Defense Objectives* (U.S. GPO, Washington, DC, 1982), 37 pp; *Civil Defense: Are Federal, State, and Local Governments Prepared for Nuclear Attack?*, Report No. LCD-76-464, 1977; *Continuity of the Federal Government in a Critical National Emergency—A Neglected Necessity*, Report No. LCD-78-409, 1978; *The Federal Emergency Management Agency's Plan for Revitalizing U.S. Civil Defense: A Review of Three Major Plan Components*, Report to the Chairman, Subcommittee on HUD/Independent Agencies, Committee on Appropriations, U.S. Senate (U.S. GPO, Washington, DC, 1984), Report No. GAO/NSIAD-84-11, 1984, 62 pp.
2. Gary K. Reynolds, *Civil Defense*, Congressional Reference Division, Issue Brief IB84128, updated 13 Nov 1984, 13 pp.
3. FEMA Reports: *U.S. Crisis Relocation Planning* (U.S. GPO, Washington, DC, 1981), Pamphlet P&P-7; *President Reagan Directs Implementation of Seven-Year Civil Defense Program* (U.S. GPO, Washington, DC 1982), News Release 82-86; FEMA, *FY85 Civil Defense Program*, statement before the Subcommittee on Strategic and Theater Nuclear Forces, Committee on Armed Services, U.S. Senate, 4 May 1984. (U.S. GPO, Washington, DC, 1984); *The FEMA Budget in Brief FY 1986* (U.S. GPO, Washington, DC 1985).
4. B. Wayne Blanchard, *American Civil, Defense 1945—1975: The Evolution of Programs and Policies*, Doctoral dissertation, Department of Government and Foreign Affairs, University of Virginia, 1979, reprinted by FEMA, 1980.
5. *The Counterfeit Ark: Crisis Relocation for Nuclear War*, edited by Langley Keyes and Jennifer Leaning (Ballinger, Cambridge, MA, 1984).
6. The Front Line, P.O. Box 1793, Sante Fe, NM 87504. (This is a newsletter critical of civil defense.)
7. "Civil Defense Relocation Plans Said to be Dropped," *The New York Times*, 4 March 1985, p. A9.
8. Robert Scheer, *With Enough Shovels: Reagan, Bush, and Nuclear War* (Vintage, New York, 1983).

Chapter 4
Effects of nuclear weapons

L. C. Shepley

Civil defense must be based on a realistic assessment of conditions during and after nuclear war. This chapter is a brief review of the effects of nuclear weapons and of what is known or surmised about conditions following a full-scale nuclear war. The standard reference on the results to be expected from nuclear explosions is by Glasstone and Dolan.[1] In fact, so much of this chapter is based on this book that we will forego specific reference to it except in special cases. Other accounts, including less technical descriptions, may be found in Ehrlich,[2] Schroeer,[3] Tsipis,[4] and von Hippel.[5] The Federal Emergency Management Agency Attack Environment Manual[6] describes the effects of nuclear explosions quite graphically. A good list of references is given by Schroeer and Dowling.[7]

There are three main ways in which nuclear weapons differ from weapons which use chemical explosives. The first, of course, is that they can release perhaps 10^5–10^7 times more energy per unit mass of weapon. The second is that such elevated temperatures are produced that a high percentage of the yield is thermal radiation. The third is the presence of nuclear radiation and radioactive fallout. The yields of nuclear weapons range from below one kiloton (kt) to many megatons (Mt). A 1 Mt bomb has an explosive power equal to that of one million tons of TNT; a 1 Mt yield is defined to be equal to 10^{15} calories (a food-type Calorie = 1000 calories) or 4.2×10^{15} J.

Nuclear weapons may be roughly divided into two classes, those used in theaters of combat (theater and tactical weapons) and those used in long-range warfare (strategic weapons). In the U.S. arsenal, tactical weapons range in yield from about 0.1 to 400 kt, strategic weapons from 40 kt to 9 Mt.[8] Many strategic weapons, including bombs and missile warheads, are within an order of magnitude of the megaton range, so we will start with a brief description of the physics of a 1 Mt weapon. (Scaling laws exist for estimating the effects of other yields; see, for example, the circular slide rule "Nuclear Bomb Effects Computer" included in Ref. 1.) Finally, we will describe the postulated effects of many nuclear explosions.

Physics of a one-megaton explosion

A 1 Mt fusion weapon has at its core a fission weapon, consisting of a sphere of plutonium which is surrounded by chemical explosives (see Fig. 1). When the chemical explosion occurs, the plutonium sphere is compressed from a non-

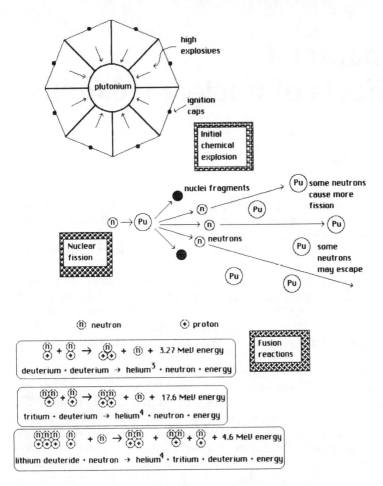

Figure 1. Composition of a 1 Mt fusion weapon. (Source: Ref. 4.)

critical to a critical mass. This critical mass is a few centimeters in radius and is the amount of fissionable material needed to sustain a chain reaction. An alternative method to start the process is to fire a shaped piece of the isotope uranium 235 (^{235}U) at another piece, again using a chemical explosion. The critical mass (the minimum amount needed for the fission process) depends on several things, including what material is used and the shape of the assembled or condensed mass. In addition, a smaller critical mass can be used if it is surrounded by materials which reflect neutrons which would otherwise escape and which thus help maintain the chain reaction. Critical masses of less than 10 kg are possible.[8]

The fission chain reaction produces lighter nuclei and extra neutrons (see Fig. 1). The excess energy which is produced appears both in the form of photons (gamma rays or γ rays) and in the kinetic energy of the fission products. The time scale for a single fission event, once a nucleus has absorbed a neutron, is the nuclear time scale of 10^{-23} s (roughly the time for light to travel across a nucleus). The time scale for a neutron traveling its mean free path (average distance between collisions) of a few centimeters is on the order of 10^{-9} s, and that for the entire nuclear fission explosion is of the order of 10^{-7}–10^{-8} s. By far the longest time scale during this initial stage is the fraction of a millisecond which is taken up in the chemical explosion.

In a fusion weapon the original ball of plutonium or uranium has several kilograms of the hydrogen isotopes deuterium and tritium nearby. A convenient way of providing this heavy hydrogen is in the form of lithium deuteride. Not only is there no need to maintain hydrogen either under pressure or at low temperatures, but neutrons react with the lithium to produce helium and tritium, the latter being used to fuel the fusion reaction (see Fig. 1). The heat of the initial fission reaction provides the kinetic energy needed to overcome the electrical potential barrier between the positively charged nuclei of deuterium and tritium, and hence the name thermonuclear weapon. The fusion phase takes place in about a millionth of a second.

The outer mantle of a 1 Mt weapon is a shell of about a hundred kilograms (or less) of the comparatively cheap and abundant isotope ^{238}U. The fast neutrons from the fusion reaction cause these uranium nuclei to fission; again the time scale is about a millionth of a second. About half the total yield comes from the fusion reaction and about half from the fissioning of the outer shell.

In this initial stage the bomb is turned into a hot, gaseous ball a few meters in diameter. This fireball consists of radioactive nuclei left over from fission reactions, helium from fusion reactions, excess neutrons, and high-energy photons (γ rays). The temperature of this initial fireball is about 10^7–10^8 K, and of course it is expanding rapidly. The energetic photons expand even more rapidly, leaving the ball at the speed of light, and heat the air surrounding the fireball. As the heat is shared by an ever increasing amount of air, the temperature of the interior of the fireball drops until it is about 3×10^5 K. The layers of air just surrounding this fireball also heat up and in so doing absorb radiation more effectively. The initial flash of light dims slightly for a brief period because of this effect; the dimming effect is less at high altitudes.

At this point the expanding bomb debris catches up with the outer edge of the heated air. The debris condenses and heats the surrounding air layers. This air forms a shock wave which leaves the fireball itself behind, a condition called "breakaway." The fireball becomes visible; its surface temperature is around 7000 K, which is slightly larger than the 6000 K temperature at the surface of the sun. It is now still under one-tenth of a second after the bomb has gone off.

An idealized observer watching from a distance would see an intense but very short flash, followed by a brief dimming at about 0.1 s, followed by the appearance of a large, expanding ball as bright as the sun. This double flash is one characteristic of a nuclear explosion. The fireball is now about 800 m in diameter and grows to a diameter of approximately 5 km in about 9 s as it begins to lift. It persists for the better part of a minute and rises as the mushroom cloud forms with a diameter of 15 km and a height of about 20 km.

While all this is happening, another effect is occurring: The photons, in

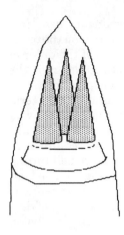

Three Mk-12A reentry vehicles inside a Minuteman III shroud

Yield: 335 kilotons each

Weight: less than 800 lb

Dimensions:
  Length: less than 6 feet
  Diameter: about 21 inches

Figure 2. Example of a nuclear weapon warhead for the Mk-12A MIRV deployed on some of the Minuteman III ICBM force. Deployment: 1083 stockpiled as of January 1983. (Source: Ref. 8.)

addition to heating the air, strip electrons from the air molecules. These electrons strike others, each initial electron creating 3×10^4 secondary electrons. The electrons are thrown outward and then rush back, attracted by the positive ions left behind (see Fig. 7). Because this oscillation is not spherically symmetric (due to inhomogeneities in the air or for weapon bursts near the ground or at high altitudes), it produces a pulse of electromagnetic radiation, called the electromagnetic pulse or EMP. The EMP can develop fields on the order of 10^4–10^5 V/m; it has a broad frequency spectrum with a maximum in the radio range around 10^5 Hz. About 0.3% of the yield appears in the form of gamma rays, and a fraction of this energy (ranging from 10^{-2} for a high altitude burst to 10^{-7} for a ground burst) appears as the EMP.

Here it is interesting to reflect on just what a thermonuclear weapon consists of. The plutonium which was involved in the initial fission reaction was only about 10 cm in size. The hundred kilograms of uranium shell and lithium deuteride take up at most only a few tenths of a cubic meter in volume. Including the chemical explosive and the detonation controls, a modern 1 Mt weapon is the size of a small trunk or large suitcase[8] (see Fig. 2). For some defense implications, see Ref. 4.

Effects of a single weapon

At this point, a few seconds after the detonation of a 1 Mt weapon, let us consider the effects on the surroundings. About 50% of the yield is in blast wave, 40%–45% is in thermal radiation, and 5%–10% is energy in the form of ionizing nuclear radiation. A tiny fraction is in the EMP (see Fig. 3).

The fireball expands to about 5 km in diameter and has the temperature of the surface of the sun. If the weapon is detonated close enough to the

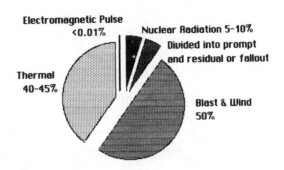

Figure 3. Effects of a 1 Mt nuclear explosion.

ground (below an altitude of about 2 km for a 1 Mt weapon) that the fireball touches earth, it is said to be ground burst. If instead it is exploded high enough that the fireball does not touch ground, the weapon is said to be air burst. A ground-burst weapon produces extreme blast effects close to its point of impact and will create radioactive fallout. An air-burst weapon produces a wider range of blast and fire effects strong enough to destroy targets which are not especially hardened, and it produces little or no fallout. Weapons may also be detonated under water, below ground, or at very high altitudes, above most of the Earth's atmosphere.

To take a specific example, we will concentrate on a 1 Mt weapon exploded right at ground level. The point of detonation or the point directly under the detonation site is called ground zero. At ground zero, soil and rock are vaporized, forming a crater (see Fig. 4). Crater depths range from 25 to 70 m, depending on the geological structure of the ground. The diameter of the visible or apparent crater ranges from 300 to 500 m. Debris is thrown out to a distance roughly twice the radius of the crater. For a radius of about 500 m around ground zero, the area suffers such complete destruction that even hardened missile silos would be rendered unusable. Much of this material is vaporized and lofted into the air to become fallout (see below). This material is made radioactive because of nuclear reactions induced by neutrons from the weapon and because radioactive weapon debris condenses with material from the crater.

Beyond the crater, the first major destructive effect comes from the thermal effects of the fireball. The intensity of the energy passing through a unit area varies approximately as the inverse square of the distance to the center of the fireball. There are many factors which must be considered. First of all, the fireball is rising and cooling, having an effective life of under a minute. Second, transfer of the heat energy depends on the clarity of the atmosphere, and absorption by the air may be quite strong. Hills and buildings will create shadowed regions. On the other hand, clouds may reflect heat energy even to regions under a direct shadow.

One can, however, give typical levels of energy transmittal for reasonably clear sky conditions and moderately flat terrain. Thus, at 12 km from ground zero, an area of a square centimeter perpendicular to the line of sight will receive about 6 calories (25 J). When this energy is absorbed in the top layer of skin on a human, it causes second-degree burns. Six cal/cm^2 is also enough to cause paper to ignite. Closer, 6 km from ground zero, construction

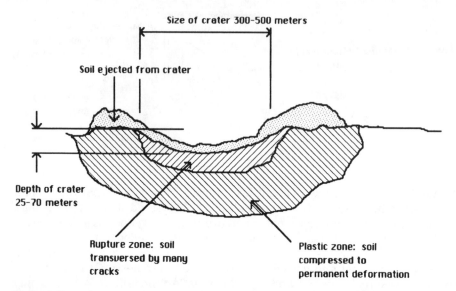

Figure 4. Crater from a 1 Mt weapon exploded at ground level. (Source: Ref. 4.)

materials such as plywood ignite. At the intermediate distance of 10 km from ground zero, the thermal energy flux is 10 cal/cm^2. This causes third-degree burns in people, the most serious class of burn injuries. It has been estimated[1] that as many as 18% of the people subjected to the total thermal output of a 1 Mt weapon within a radius of about 15 km from ground zero will receive second- or third-degree burns, the rest receiving first-degree burns. These burns will occur in clear weather conditions, on exposed skin, for people caught in the open.

A 1 Mt weapon will yield 50% of its thermal energy in 2 s; a 10 kt weapon expends 50% of its thermal energy in about $\frac{1}{4}$ s; and a 10 Mt weapon takes over 5 s to yield 50% of its thermal energy.[1] Thus, the thermal energy of a 1 Mt explosion is emitted over a long enough period that some people, on seeing the start of the fireball, will be able to find some cover; in contrast, the thermal energy from weapons of 100 kt or less appears too quickly for people to react. Also, the effect of a given amount of energy in causing a particular class of burn varies with the rapidity of absorption, the slower rate of large weapons meaning that more energy must be absorbed to cause a specific severity of burn.

Within the 10 km radius much that is burnable will catch fire, and major fires will start. The blast and wind effects described below tend to blow these fires out, but they also start other fires from ruptured gas lines or broken fuel storage tanks. Under some circumstances, not well understood, a firestorm could start within this 10 km radius. A firestorm is a several-hour-long fire which is so strong that its updrafts create a chimney effect. The term "firestorm" is meant to contrast with a conflagration, or large moving fire. The term "group fire" is also used[6] for a large fire of lower severity than a firestorm; the fire at Hiroshima would then be classified as a group fire rather than a firestorm. Winds of perhaps 200 km/h come in from all sides to feed the central cylinder of uprushing air. As the fire progresses, it does not spread but instead consumes everything combustible within its area. Central temperatures can reach thousands of degrees. The devastation of Hiroshima was in great part due to the fire there.

Experience in World War II showed that it is uncertain what conditions are necessary to start a firestorm. In general, it is estimated[1] that a loading of about 4 g/cm^2 of burnable material is needed for a firestorm. In addition, at least half of the structures must be on fire simultaneously, the wind must be less than 13 km/h, and the burning area must be over a square kilometer. There are many uncertainties, for example, as to what will burn in a modern city (most likely the interiors rather than the exteriors of buildings will burn) and whether high-rise buildings would inhibit firestorms.[1] Whether many fires would start in a nuclear war or whether few would start is an important but unanswered question. The soot created by a large number of fires could have disastrous effects on climate, and this aspect is discussed in detail in Chap. 9.

In addition, the thermal effects of a nuclear weapon cause nitrogen oxides to be formed in the atmosphere. The rising fireball injects them into the stratosphere where they combine with ozone, destroying the ozone layer which shields the Earth's surface from harmful ultraviolet radiation.

The next effect is due to the blast caused by the outward expansion of the fireball. The leading edge of the blast shock wave travels out from ground zero at speeds just above the speed of sound, about 350 m/s. The overpressure in the shock front is often given in lb/in.^{2}, or psi, or in decibars (1 psi is about 0.7 decibar or 0.07 bar) and should be compared with ambient air pressure of 15 psi (about 1 bar or 10 decibar). Of course the highest overpressures occur near ground zero, but military targets may be hardened even against these. The most hardened targets, such as missile silos buried in the ground, will also receive shocks transmitted through the earth, and the degree of damage depends to some extent on how hard these silos are made. In general, it has been estimated[4] that these targets will be rendered useless only if they are within the cratering area.

Outside the cratering region, the overpressure decreases with distance. Following the leading edge of the shock wave is a wind which first rushes outwards from ground zero, then back. These afterwinds directed toward the center of the explosion are caused by the updrafts which accompany the rising of the fireball. The strength of the wind is roughly a function of the shock overpressure (see Fig. 5). The effects of this wind are ascribed to "dynamic overpressure," but in practice the effects of the dynamic overpressure are usually lumped together with those of the overpressure in the shock. Thus when one speaks of the blast effects at a distance where the overpressure is 5 psi (3 decibar), one also includes the results of the winds following it. These winds reach a velocity of over 200 km/h, with a diminished speed for the succeeding inward winds. An overpressure of 5 psi (3 decibar) causes tremendous forces on the sides of buildings, where they are not able to support sudden forces. At 5 psi (3 decibar) even brick and cinder-block walls can collapse. This pressure is the threshold pressure for eardrum rupture, though it takes higher overpressures to cause other significant direct injuries in people. Indi-

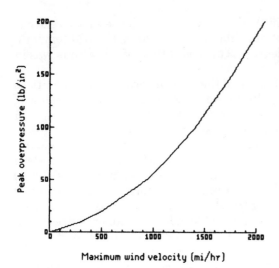

Figure 5. Relation between peak overpressure and maximum wind velocity calculated for an ideal shock front. (Source: Ref. 1.)

rect injuries, for example, injuries due to flying objects, can be severe even at lower overpressures. Glass windows shatter even at overpressures of under 1 psi (0.7 decibar).

For a 1 Mt weapon burst on the ground, the overpressure is 200 psi (13 bars) at 800 m from ground zero, and the shock is followed by 3000 km/h winds. At about 2500 m from ground zero, the overpressure is 15 psi (1 bar) and the winds reach 600 km/h. At 5 km the overpressure is 5 psi (3 decibar) and the winds over 200 km/h (see Fig. 6). Thus at a distance of 5 km from ground zero brick houses and concrete or cinderblock walls are usually destroyed, and the area of destruction is 80 km^2 (30 mi^2). A person within this area has a 50% chance of surviving.

For any given overpressure (such as 5 psi or 3 decibar), there is an optimal height at which the range of the overpressure is greatest. A 1 Mt weapon detonated about 3 km above the ground will extend the 5 psi (3 decibar) range to a range of between 6 and 7 km, although no crater would be formed. In such an air burst, reflected waves from the ground add to the direct shock, producing enhanced shock effects. For a city or an industrial complex which does not have extremely hardened targets, an air burst would be used; the area of destruction being about 130 km^2 (50 miles^2).

The third major portion of the yield, 5%–10%, is in the form of nuclear radiation. The ionizing properties of this radiation can cause sickness or death. The measure of exposure is in rems (roentgens equivalent for man, units which take into account the differences in the biological effects of various types of radiation). A dose of 100 rems or less causes a cutback in white cell production, leading to increased susceptibility to infection. A dose of 150 to 200 rems causes loss of hair and nausea, and a small percentage of people will die. An exposure to 450 rems kills 50% of the people. Higher exposures kill a greater percentage of those exposed; a dose of between 600 and 1000 rems will cause 90% to 100% deaths over a period of one to six weeks, An extreme dose of several thousand rems is needed to produce death very

Effects of nuclear weapons 55

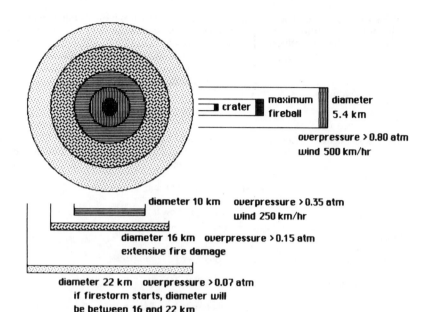

Figure 6. Blast and thermal effects of a ground-burst 1 Mt weapon. Within the 10 km diameter circle there is almost complete destruction with 50% dead, 40% wounded. Within a 19 km diameter circle thermal effects can cause third-degree burns.

quickly, that is within a couple of days. The general clinical effects of nuclear radiation are given in Table I.

Death from a barely lethal dose of nuclear radiation comes after a period of several weeks or even months. The distance from ground zero for a lethal exposure to prompt nuclear radiation is a bit over 2 km for a 1 Mt weapon. Thus, a person is much more likely to be killed by heat or blast effects than by nuclear radiation effects. In a smaller weapon, the nuclear radiation effects are relatively more important. The bombs used on Hiroshima and Nagasaki did cause significant amounts of nuclear radiation sickness (though most immediate casualties were due to blast and fire). Small tactical nuclear weapons, for example, some of those designed to be used by troops in the field, also would result in nuclear radiation effects of significance. Enhanced radiation weapons, or the so-called neutron bombs, are those in which the yield in radiation is increased relative to the blast and heat effects. This enhancement is militarily useful in very small weapons, about 1 kt in yield,[5] and it is produced by enhancing the fission in such a weapon with deuterium-tritium fusion reactions, these being more efficient at producing excess neutrons.

Nuclear radiation also appears in the form of fallout. For civil defense considerations, this source of radiation exposure to people would be more significant than would direct exposure (except possibly for people in blast shelters). We will return to the effects of radioactive fallout, for these effects cover a time period that is long in comparison with the prompt effects so far described.

Table I. Clinical effects of nuclear radiation. The effects of a dosage of 0–100 rems are generally not noticeable. Long-term effects such as increased likelihood of cancer and genetic repercussions are not included. The generally accepted dosage for a 50% death rate is 450 rems. (Source: Ref. 1.)

Dose in rems	100–200	200–600	600–1000	over 1000
Incidence of vomiting	Sometimes	Often	100%	100%
Initial phase	3–6 hours	0.5–6 hours	0.2–0.5 hour	5–30 minutes
Characteristic signs	Decline in white cells	Severe decline in white cells; loss of hair; hemorrhage		Diarrhea; fever; possible convulsions
Therapy	Little needed	Blood transfusion; antibiotics; consider bone marrow transplant		Maintain fluid; give sedatives
Prognosis	Excellent	Guarded	Guarded	Hopeless
Incidence of death	None	0–90%	90–100%	100%
Death occurs within	...	2–12 weeks	1–6 weeks	1–14 days
Cause of death	...	hemorrhage; infection		circulatory or respiratory failure

The last prompt effect we will discuss (although it is one of the first effects to occur after detonation) is the electromagnetic pulse or EMP (see Fig. 7). Because of the EMP, a 1 m wire or other conductor, acting as an antenna, can develop voltages of over 10^4 V. Sensitive equipment connected to wires can be affected to great distances. Telephones, life-support systems in hospitals, even street light systems, can be damaged unless they are protected or hardened. A 1 Mt ground burst can cause damage to unhardened electrical equipment to a distance of 20 km, though the effect falls off rapidly. The spectacle of all cars coming to a halt because their electronic ignitions have blown out, however, may not happen: On the one hand, metal car bodies and the metal of the engines tend to act as shields; on the other hand, electronic devices used in newer cars are quite sensitive to a pulse of radiation of this sort.

The most spectacular effects of the EMP come from a very high altitude burst, one which occurs above most of the atmosphere, say 50–300 km above the Earth's surface. The photons from the explosion have no air to impede their motion until they strike the atmosphere and generate the EMP. The deposition region can occur in a volume of atmosphere 1000 km in radius and averaging about 50 km in thickness (see Fig. 7). The EMP for a high altitude burst is strong throughout the area in the line of sight of the explosion, and it is possible that this EMP could cause a massive electric blackout and incapacitate communications over much of the U.S.

Before turning to the effects of many nuclear detonations, let us summarize the effects of a single ground-burst weapon. The crater produced by a 1 Mt

Effects of nuclear weapons

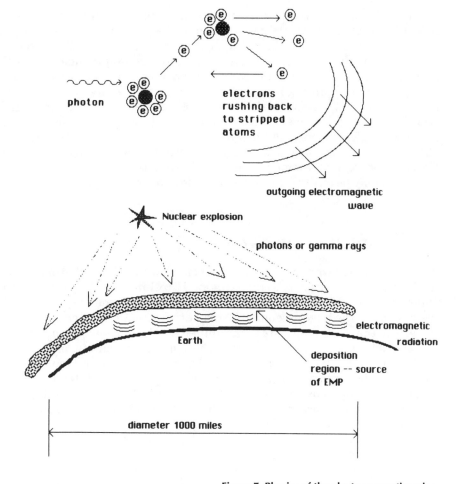

Figure 7. Physics of the electromagnetic pulse.

weapon exploded on the ground extends to a diameter of 300 to 500 m. The major thermal effects occur to a radius of 10 km. Within this radius a firestorm could occur. Out to a radius of 5 km, the overpressure in the shock front reaches 5 psi (3 decibar). It is out to this radius that the major blast effects occur against structures which are not especially hardened. A lethal dose of prompt nuclear radiation occurs to a radius of about 2 km. EMP effects from a ground burst occur to perhaps 20 km.

Effects due to air burst weapons differ somewhat in that no cratering occurs, and the blast effects against city structures cover a larger area. A weapon burst above the atmosphere causes no blast or radiation effects but does cause a very wide ranging EMP, as mentioned above. Weapons exploded beneath the ocean or below ground also have unique effects (for example, waveform signatures which are useful in detection), but we will not discuss them here.

The effects of larger or smaller weapons may be estimated by scaling. The size of the fireball is basically a function of the yield Y, and the volume is roughly proportional to Y; the radius actually varies as $Y^{0.4}$.[1] The range of a given blast or thermal effect scales, to a reasonable approximation, as the cube root of Y. The effects of nuclear radiation, however, reach a range due to the interaction of ionizing radiation with air. Consequently, the relative importance of nuclear radiation rises as the yield decreases, and enhanced radiation weapons of 1 kt yield have been designed for tactical use. With the larger strategic weapons, the major direct damage is caused by blast and heat. The majority of such weapons have yields within a factor of 5 or 10 of 1 Mt. The cube root scaling law means that the effects of these weapons are close to, or at least may be easily derived from, those of a 1 Mt weapon.

Effects of nuclear war

The only nuclear war in history was decided when the U.S. dropped two bombs, each small by today's standards, on Hiroshima and Nagasaki. A very limited war involving only a couple of strategic weapons may not now be possible, given the huge arsenals of the superpowers. As far as civil defense is concerned, the consequences of a war limited to a few nuclear detonations are the same as those of a single weapon. There is, unfortunately, incentive for escalation of a nuclear war, once started, to a full-scale war. One incentive comes from the rapid pace of a nuclear conflict: A land-based intercontinental ballistic missile needs but 30 minutes to reach its target; a submarine-launched missile typically needs much less time, about 5 to 7 minutes. In addition, the debris carried up by the mushroom cloud, along with the great turbulence created by one large-scale nuclear weapon affects unpredictably the trajectory of a second missile for perhaps an hour afterward. This "fratricide" effect is a secondary inducement to wage a war as suddenly and as thoroughly as possible. And needless to say, the most difficult incentive to assess is the natural military tendency that once the enemy is engaged, one proceeds with as much strength as possible. In describing the effects of nuclear war, we will therefore concentrate on the scenario of a full-scale attack. For comments on waging war, see Dunnigan,[9] Powers,[10] and Tsipis.[4] The study by the Office of Technology Assessment[11] also includes a discussion of war scenarios.

A counterforce attack is one primarily involving military targets. In the U.S. there are just over 1000 missile silos which would be targeted in such a case. For each silo, it is likely that two 1 Mt weapons would be used, one air burst to maximize general blast effects and the second ground burst to make a crater large enough to prevent the missile from being launched. Airfields and submarine bases would be hit by another 150 weapons. Thus even a counterforce attack on the U.S. would involve explosions of 1000 Mt of weapons. A counter-recovery attack is an attack on cities and industrial capacity and is meant to destroy a country thoroughly. To destroy 80% of our manufacturing capacity would also require over 400 1 Mt weapons against major industrial complexes and almost 300 smaller weapons against such specific targets as

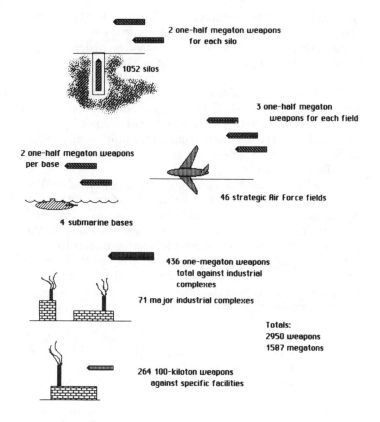

Figure 8. Counterrecovery attack against the U.S. (Source: Ref. 4.)

refineries, mills, and other factories.[11] An attack by us on the U.S.S.R. would also use upwards of 1500 megatons (see Fig. 8). It should be noted that even in such a scenario, the U.S. and the U.S.S.R. would each use less than half its nuclear arsenal.

The effects of such a full-scale attack are not simply the sum of the effects of the individual weapons used. To take one important example, the overlap of weapons detonated near each other produces overkill without enlarging the area of destruction greatly. The total area of fire and building destruction will be less than the sum of the areas of the weapons taken one by one. Even if the weapons do not overlap greatly, a given megatonnage deployed will destroy a greater total area if it results from several smaller weapons rather than one large one, because of the one-third power scaling law discussed above. Thus, let M be the total megatonnage deployed, and let n be the number of weapons, so that M/n is the yield per weapon. The radius of destruction per weapon is roughly proportional to $(M/n)^{1/3}$; the area of destruction per weapon is proportional to $(M/n)^{2/3}$; and the total area of destruction for n nonoverlapping weapons is therefore proportional to $n^{1/3}$. This effect, that the smaller the yield per weapon (and therefore the larger the number n used in a given total deployment of M megatons) is the cause of the "effective megatonnage" measure of an arsenal, a measure which weights smaller weapons somewhat more heavily per megaton than larger ones. The effective megatonnage or megatonnage equivalent of a weapon having Y megatons in yield is $Y^{2/3}$.[5] The overlapping and scaling effects, however, are simply a matter of looking at the detailed effects of a given deployment pattern. There are three effects from the use of a large number of weapons that can be extremely dire and qualitatively different from a summation of heat and blast. These are radioactive fallout, effects on the ozone layer, and climatic effects, the so-called nuclear winter.

The fission reactions produce radioactive nuclei. When the weapon is air burst, so that the fireball does not touch the ground before it begins to rise into the atmosphere, these radioactive nuclei are carried to high elevations, to about 20 km (above the tropopause at about 9 km, into the stratosphere) for a 1 Mt weapon. Smaller weapons loft these nuclei to lower elevations, but in general the lofting is enough so that they spread and only slowly settle to earth after much of the radioactivity has dissipated. There will be statistically significant increases in cancer and genetic troubles, but these effects are minor in comparison with the initial destruction caused by air-burst weapons.

If the fireball touches the ground, the soil and rock will become radioactive. Much of the soil material is vaporized but then condenses into fine particles. How much fallout would be produced is highly uncertain, for nuclear weapons tests have not involved a full variety of soil types. This dust is light enough to be carried by surface winds and yet sufficiently dense so as to settle downwind. Depending on the wind at the time of detonation, it may be carried for hundreds of kilometers. About 50% to 70% of the radioactive dust from a ground-burst weapon returns to ground in about a day. An accumulated dose larger than 100 rems after one week, with a 25 km/h wind, from a 1 Mt weapon, would cover over 5000 km^2 or an oval extending downwind almost 300 km. The area of contamination is not by any means spread in a uniform pattern. Moreover, surface winds alone cannot be used to gauge the pattern of fallout, since upper level winds can play an important part. The unevenness of a realistic fallout pattern is often not sufficiently emphasized.[1] Furthermore, the dust, although it may include quite visible particles, is not visually recognizable as being radioactive. Instruments must be used to tell reliably where hot spots of high dose rates occur.

The overlapping areas of contamination from a full-scale attack would blanket most of the country, since our military bases and industrial centers are widespread. Fallout protection is not a simple matter of obtaining shelter, and the problems of sheltering and monitoring fallout are treated in Chaps. 5 and 6 on radiation and sheltering. Radioactive contamination consists of a variety of nuclides with a variety of half-lives, so that a calculation of the attenuation of radioactivity is complex. The "7-10 rule" is used; it says that the radiation intensity is reduced by a factor of 10 for each factor of 7 increase in the time since detonation.[6]

Current standards say that a safe level of radiation for permanent occu-

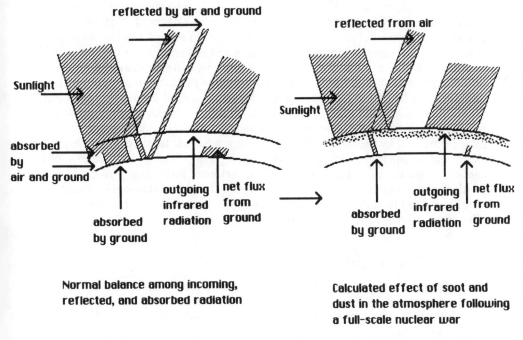

Figure 9. Nuclear winter. (Source: Ref. 13.)

pation would be a dose rate of 2 rems/year. The area which would fail to meet this safety standard for 2 months because of a 1 Mt, ground-burst weapon, if the wind can be taken as a steady 25 km/h, covers 4×10^4 km². Since the area of the continental U.S. is 8×10^6 km², a full-scale war, in which thousands of weapons are used, would thus render a large part of the U.S. subject to many times the level currently considered safe. Even such elevated levels of radiation would not necessarily lead to immediate life-threatening situations, but it is clear that a drastic relaxation of present radiation standards would have to be made for major parts of the U.S. to be considered habitable in the months following a nuclear war.

It was mentioned earlier that the heat from a thermonuclear weapon creates nitrogen oxides in the atmosphere and then carries these oxides to high levels. At high atmospheric levels, these oxides react with ozone and destroy it.[11] The ozone level could be so seriously depleted that it might be several years before it could reform. There is much that is unknown about the effects on the ozone layer. In the first place there are large natural variations in the ozone layer. It is not known just how effective the nitrogen oxides would be in destroying ozone. Finally, the level of lofting depends on the details of what size weapons are used. A 1 Mt or larger weapon creates a cloud which reaches 20 km into the atmosphere. A height of 24 km is needed for significant ozone depletion.[11] Smaller weapons do not reach as high, but larger ones do reach higher. The ozone problem is discussed further in Chap. 10.

The third and potentially most important of the global effects is what is known as nuclear winter (Peterson,[12] Turco et al.,[13] Ehrlich et al.[14], Levi and Rothman,[15] and Committee on the Atmospheric Effects of Nuclear Explosions[16]). The causes of climatic effects are dust and soot. A nuclear explosion would produce massive amounts of dust, and fires from the weapons would produce massive amounts of soot.[17] Dust mainly scatters sunlight, letting a lot through, and so does not change global weather greatly. However, the soot generated by firestorms in cities and in forests is a good absorber of sunlight. Much of the absorbed sunlight is then reradiated to outer space, so that the net effect is to decrease the energy reaching the ground. At the same time, the ground cools off by radiating infrared energy, which is let through by the soot cloud and escapes to outer space (see Fig. 9). The effect on the interior of a continent would be rapid cooling. The darkness following a nuclear war would last as long as the cloud of soot stays in the upper atmosphere, and the resulting drop in temperature could be down to well below freezing. These effects are discussed further in Chap. 9.

Conclusions

The immediate effects of a nuclear weapon take place over a time scale of at most a few minutes. These include blast, heat, and prompt nuclear effects. These prompt effects can be fairly well predicted, though details of topography and city construction make predictions quite difficult. On a time scale of one hour to one day, the effects leading to firestorms and fallout are much more uncertain. It is sure that fallout will occur and that firestorms will occur in at least a small percentage of the cases, but it cannot be predicted which cities will burn and where fallout will be concentrated.

Global and long-time effects are even more unforeseeable. Nobody can say for sure what the effects on the ozone layer will be, or how these effects will be felt by agriculture and animal life. Nuclear winter is not a certainty following a full-scale war, but it is a possibility. Even the possibility could affect the way nuclear war planning is carried out. It has been reported that a future emphasis may be on a large number of very small weapons rather than a few thousand megaton-level weapons. This might decrease the number of large fires and lessen the likelihood of a nuclear winter.[10] (There seems to be a trend toward smaller weapons for other reasons, anyway.)

The implications for civil defense of the possibility of nuclear war are of course tremendous, for it is likely that destruction will fall on participant and nonparticipant alike. Although many of the effects of nuclear weapons are known, it is clear that many others, particularly those that can affect civil defense planning, can only be guessed at. These global effects, however, are well enough understood that many would agree that civil defense plans cannot begin to be effective to counteract the worst possibilities. However, it would be wrong to let the worst possible case blind one to making reasonable civil defense plans for various other scenarios, in which only some of the effects described here occur.

References

1. S. Glasstone and P. J. Dolan, *The Effects of Nuclear Weapons*, 3rd ed., edited by S. Glasstone and P. J. Dolan (U.S. Departments of Defense and Energy, U.S. GPO, Washington, DC, 1977).
2. R. Ehrlich, *Waging Nuclear Peace: The Technology and Politics of Nuclear Weapons* (State University of New York Press, Albany, 1985).
3. D. Schroeer, *Science, Technology, and the Nuclear Arms Race* (Wiley, New York, 1984).
4. K. Tsipis, *Arsenal: Understanding Weapons in the Nuclear Age* (Simon and Schuster, New York, 1983).
5. *Physics, Technology, and the Nuclear Arms Race (APS, Baltimore, 1983)*, edited by D. W. Hafemeister and D. Schroeer, AIP Conf. Proc. No. 104 (American Institute of Physics, New York, 1983). See especially the article by F. von Hippel, pp. 1–46.
6. Federal Emergency Management Agency (FEMA), *FEMA Attack Environment Manual*, Pamphlets No. CPG2-1A1, No. CPG2-1A2, No. CPG2-1A3, No. CPG2-1A4, No. CPG2-1A5, No. CPG2-1A6, No. CPG2-1A7, No. CPG2-1A8, No. CPG2-1A9, May, 1982 (reprint with minor changes from *DCPA Attack Environment Manual*, June 1973 (FEMA, Washington, DC, 1982). Chap. 1: "Introduction to Nuclear Emergency Operations." Chap. 2: "What the Planner Needs to know about Blast and Shock." Chap. 3: "What the Planner Needs to Know about Fire Ignition and Spread." Chap. 4: "What the Planner Needs to Know about Electromagnetic Pulse." Chap. 5: "What the Planner Needs to Know about Initial Nuclear Radiation." Chap. 6: "What the Planner Needs to Know about Fallout." Chap. 7: "What the Planner Needs to Know about the Shelter Environment." Chap. 8: "What the Planner Needs to Know about the Post-Shelter Environment." Chap. 9: "Application to Emergency Operations Planning."
7. D. Schroeer and J. Dowling, Am. J. Phys. **50**, 786 (1982).
8. T. B. Cochran, W. M. Arkin, and M. M. Hoenig, *Nuclear Weapons Databook: Vol. I, U.S. Nuclear Forces and Capabilities* (Ballinger, Cambridge, MA, 1984), first of an eight-volume set.
9. J. F. Dunnigan, *How to Make War—A Comprehensive Guide to Modern Warfare* (Quill, New York, 1983).
10. T. Powers, The Atlantic Monthly **254** (No. 5), 53 (1984).
11. Office of Technology Assessment (U.S. Congress), *The Effects of Nuclear War* (Allanheld, Osmun & Co., Totowa, NJ, 1980), issued by the Office of Technology Assessment, Report No. OTA-NS-89, 1979.
12. *The Aftermath: The Human and Ecological Consequences of Nuclear War*, edited by J. Peterson (Pantheon, New York, 1983).
13. R. P. Turco, O. B. Toon, T. P. Ackerman, J. B. Pollack, and C. Sagan, Science **222**, 1283 (1983); Sci. Am. **251**, (No. 2) 33 (1984).
14. P. R. Ehrlich, C. Saan, D. Kennedy, and W. O. Roberts, *The Cold and the Dark—The World after Nuclear War* (Norton, New York, 1984.)
15. T. Rothman and B. Levi, Phys. Today **38** (No. 9), 58 (1985).
16. Committee on the Atmospheric Effects of Nuclear Explosions, G. F. Carrier, Chairman, Commission on Physical Sciences, Mathematics, and Resources, National Research Council, *The Effects on the Atmosphere of a Major Nuclear Exchange* (National Academy Press, Washington, DC, 1985).
17. A. A. Broyles, Am. J. Phys. **53**, 323 (1985).

Chapter 5
Nuclear radiation and fallout

Cliff Castle

Background on radiation

The advent of nuclear weapons has produced totally new problems for protection during and after an attack. Before nuclear weapons, the necessity was to protect against blast, which covers a relatively short period of time. Nuclear radiations produced from a nuclear detonation now force planners to provide protection for weeks after the attack.

These radiations consist of alpha particles (nuclei of helium atoms), beta particles [electrons (negative or positive)], gamma rays (high-energy photons), and neutrons. Most of the neutrons are emitted at the time of the explosion during fission or fusion. Some of the gamma rays are produced at the same time while the rest of the gamma rays are emitted along with alpha and beta particles during the radioactive decay of fission products, fusion products, and unfissioned uranium or plutonium. Another important source of gamma rays is neutron capture by elements (especially nitrogen) in the atmosphere and ground.

It is customary to separate this radiation into two parts—that emitted within one minute around the burst point and that caused by the return of radioactive particles (fallout) over an extended period of time to the ground. The former is called prompt or initial nuclear radiation and any radiation after one minute is considered as residual. The time limit of one minute is rather arbitrary but is based on the fact that gamma rays from a 20 kt explosion have a range (in air) of approximately 3.2 km (2 miles). Since it takes about one minute for a 20 kt cloud to rise to an altitude of 3.2 km, this was determined by weapons experts to be the time of the initial radiation. After one minute any radiation emitted from the cloud would reach the earth in insignificant amounts.[1]

The initial nuclear radiation, most of which will be released in the first seconds following a nuclear explosion, consists essentially of fast neutrons and gamma rays. Although part of the radiation will be absorbed by the weapon material itself, large amounts will escape to represent a significant hazard to living organisms and sensitive electronic equipment. The neutrons will also induce some activity in the soil around ground zero. The amount of this activity is determined by the chemical composition of the soil.

Because radiation is attenuated rather rapidly in air, distance from

Table I. Relationship of blast and initial nuclear radiation[a] (near-surface bursts[b]).

Blast overpressure (psi)	Kilometers from ground zero			
	10 kt	40 kt	100 kt	1 Mt
1	2.9	4.5	6.3	13.7
6	1.0	1.6	2.1	4.5
20	0.5	0.8	1.1	2.2
Initial nuclear radiation (rem)	Kilometers from ground zero			
	10 kt	40 kt	100 kt	1 Mt
100	1.6	1.9	2.1	3.1
500	1.3	1.6	1.8	2.6
3000	1.0	1.3	1.4	2.2

[a] Thermonuclear detonation with 50% fission.
[b] Height of bursts: 10 kt, 131 m; 40 kt, 207 m; 100 kt, 283 m; 1 Mt, 610 m.

ground zero and height of burst are important factors in determining the initial dose of radiation. In general, two points approximately 400–600 m (0.25–0.4 mile) apart will have a factor of 10 difference in initial radiation exposure. Therefore 800–1200 m (0.5–0.8 mile) of air will generally reduce the exposure by a factor of 100. As the height of burst increases from a low air burst to a high air burst, the amount of radiation reaching the ground diminishes rapidly. A 1 Mt burst at an altitude of 5 km (3 miles) would give a negligible dose at ground zero and so would a 100 kt burst at an altitude of roughly 4 km (2.5 miles).

As guidance systems for missiles have become more accurate both the U.S. and the U.S.S.R. have decreased the yield of many of their warheads. This is known as downsizing and has added further complications for a planner to protect the population from initial nuclear radiation (INR). Table I shows that as yield decreases initial radiation begins to outrange blast effects for active protection measures. For example, an unsheltered population 2.4 km from ground zero for a 1 Mt weapon would certainly receive a lethal dose of radiation immediately. However, death would occur instantaneously from the blast rather than the approximate two-week elapsed time due to radiation exposure. A population the same distance from a 100 kt explosion would experience a 50% death rate due to the blast effects. Of those who survive approximately 5% would suffer acute effects of radiation exposure (three hours for onset of vomiting). Without proper medical care a small percentage of those surviving the blast would be expected to die within two months due to the complications of radiation exposure. For a 40 kt explosion few very serious injuries would be expected from the blast. Within the first minute the entire population of a 21 km² area would receive the maximum allowable annual radiation exposure as outlined in present federal guidelines. The significance of this fact points out the tremendous difference between peacetime and wartime radiation exposure. Federal guidelines set an upper limit of 0.5 rem exposure to the general population over the course of a year. During a nuclear attack this limit will be exceeded within a matter of seconds to weeks which means that within the population an increase in the numbers of cancers and genetic problems will be expected. Although gamma rays and neutrons travel in straight lines, collisions with air molecules will cause the radiation to reach a location from all directions. This phenomenon is referred to as "skyshine" and implies that the only acceptable shelter is one that completely surrounds an individual. Protection against radiation is accomplished by either scattering or absorbing the radiation so that it does not penetrate the shelter. Three important types of interactions of gamma rays and matter are the Compton effect, photoelectric effect, and pair production. In the Compton effect a photon undergoes an elastic collision with an electron and bounces off in a different direction with less energy. The photoelectric effect involves a transfer of energy from the photon to an atom; the photon disappears and the electron is removed from the atom, thus producing an ion pair. Pair production is the creation of an electron and positron when a photon of minimum energy of about one million electron volts (MeV) passes near the nucleus of an atom. Pair production annihilates the gamma ray, as does the photoelectric effect. In all three types of interaction the probability of a photon interacting with an atom increases with increasing atomic number. Therefore, materials containing elements of high atomic number are preferred as shielding from gamma rays.

In most physics texts the equation

$$I = I_0 \exp(-\mu x) \qquad (1)$$

is used to describe the attenuation of gamma rays passing through a material. I_0 is the initial intensity while I is the intensity after passing through some thickness x of an absorber. The linear attenuation coefficient μ is determined by the material and the energy of the photons. Equation (1) holds only for a collimated, monoenergetic beam of gamma rays with the assumption that the photons undergoing Compton scattering are removed from the beam. This last assumption is true only for thin shields. In thick shields, such as needed to protect the population during nuclear attack, photons can undergo multiple Compton scattering and be directed back into the beam. Equation (1) is then modified to include a "buildup factor" and becomes

$$I = I_0 B(x) \exp(-\mu x), \qquad (2)$$

where the value of $B(x)$ is determined by the thickness of the material, the nature of the material, and the energy of the photons. Theoretical calculations show that the values for $B(x)$ can range from 10 to 100.[1] Therefore, a simple application of Eq. (1) can lead to gross errors in determining the protection afforded by various materials.

Even the use of Eq. (2) can lead to serious errors since it still pertains to monoenergetic, collimated beams. Practical radiation shelters must not only deal with a broadened, polyenergetic flux of gamma rays, but in the case of initial nuclear radiation, absorption of neutrons might also be a requirement.

Shielding against neutrons is far more problematical than gamma-ray shielding. Since Eqs. (1) and (2) are exponential in material thickness, a sufficiently thick shield will attenuate almost all gamma rays. This simple relation does not hold for neutrons because the processes of attenuation of neutrons are different from those of gamma rays. Neutrons have energies ranging from about 10 MeV for fission weapons and about 14 MeV for thermonuclear weapons down to a lower end of a few tenths of an electron volt.

Neutrons undergo very little interaction with the electrons of an atom and must interact with nuclei if they are to be removed from the initial radiation. As with gamma radiation the neutrons must either be scattered or absorbed. "Slow" (low-energy) neutrons are most likely to be absorbed. Slow neutrons are commonly termed thermal neutrons and have energies of approximately $\frac{1}{40}$ eV. "Fast" (high-energy) neutrons must thus be brought down to the slow (thermal) range for effective absorption. This reduction in energy is brought about by scattering, and the amount of energy reduction is determined by the mass of the scatterer.

For neutrons, scattering can be either elastic or inelastic. When a neutron undergoes inelastic scattering, part of its energy goes into exciting the target nucleus to a higher energy level. The scattered neutron now has less energy and the excited nucleus emits a gamma ray (or rays) as it decays back to its ground state. In order for inelastic scattering to occur, the neutron must have at least the excitation energy of the target nucleus. This means that the probability of inelastic scattering occurring is determined by the target isotope and the neutron's energy. A rough rule of thumb is that many moderately heavy (including iron) and heavy nuclei will inelastically scatter neutrons with energies ranging from a few tens to a few hundred keV. Inelastic scattering occurs for lighter nuclei only when the neutrons are more energetic.

In elastic scattering there is a simple exchange of kinetic energy from the neutron to the target nucleus. The target nucleus undergoes no excitation and so there is no gamma radiation that follows elastic scattering. Since the excitation energy is of no consequence, elastic scattering can take place with any nucleus and a neutron with any energy. Since the scattering is elastic, those elements of low mass number are most efficient for rapidly degrading fast neutrons. Those compounds containing large quantities of hydrogen are best, but because of their low densities and large thicknesses required for appreciable neutron attenuation, inelastic scattering of neutrons of energies greater than 0.5 MeV is more advantageous.

Once neutrons are slowed to the thermal range they are most likely to be absorbed. The probability of neutron capture, for any isotope, is a function of neutron energy which usually varies in a very complex manner. Hydrogen, boron, and cadmium are three common elements that are good for neutron absorption.

Concrete, a very common material, represents a good compromise for reducing the neutron flux. Although concrete does not normally contain elements of high atomic weight, it does have a fairly large proportion of hydrogen to slow down and capture neutrons. Thirty centimeters (12 in.) of concrete will decrease the neutron flux by about a factor of 10 while 61 cm gives a reduction of approximately 100. A modified concrete can give good protection from neutron radiation. An 18 cm thickness of concrete mixed with large proportions of barytes (barium), limonite (iron), or colemanite (boron), and incorporating small pieces of iron or steel can reduce the neutron flux by a factor of approximately 10.[1]

A measure of the protection afforded against initial nuclear radiation (INR) is given in terms of an INR protection factor which is the ratio of the dose in the open to the dose in the location described, at the same distance from the detonation. Table II gives the INR protection factor for various structures in relation to their desirability as blast shelters, A being the most desirable and I the least. The lower number shown relates to locations near entrances and windows and the higher number pertains to locations remote from such openings.

Fallout

Although the initial nuclear radiation is an immediate problem, the formation of fallout requires protection of the population for periods of weeks to months. Fallout is usually placed into two categories: early (local) and delayed (long-range) fallout. Early fallout is that deposited on the ground in the first 24 hours after detonation. This time limit is used since the fallout pattern is pretty well established after 24 hours from a single detonation.

The first thing that appears after the detonation of a nuclear weapon is a roughly spherical, highly luminous ball of gases called the fireball. Because of the extremely high temperatures everything in the immediate vicinity is vaporized: the fusion and/or fission products, the uranium or plutonium which has escaped fission, the bomb casing and other weapon parts. For a surface burst, a considerable amount of rock, soil, and other material will be vaporized and taken into the fireball.

As the rising fireball cools, material entering later is only melted, and as the fireball cools further and forms the mushroom cloud, some material reaching the cloud level is virtually unchanged. Inside the mushroom cloud there is a violent, toroidal circulatory motion which allows mixing of the radioactive residues with the soil components. As the fireball cools below the boiling point of the vaporized soil material, it begins to condense into liquid droplets, which eventually solidify into glasslike particles. A small proportion of the solid particles formed are contaminated fairly uniformly throughout with the radioactive residues, but in the majority, the contamination is found mainly in a thin shell near the surface.[3]

This is because as the fireball cools, the first major step in the formation of fallout is the condensation of vaporized soil. Then those radioactive elements that have already condensed are readily incorporated into the liquid droplets. The more volatile radioisotopes are still gaseous and not available to be mixed with the liquid soil droplets. Some radioactive elements do not interact significantly until the bulk of the material has cooled sufficiently to solidify. They then tend to lodge on the surface of the fallout particle and form a thin shell of radioactive contamination. This alteration in the composition of fallout is termed fractionation.

Fallout particles are quite small but under most conditions where a significant radiation hazard exists the fallout will be visible. A peak dose rate of 0.5 roentgen per hour is the level above which the Federal Emergency Management Agency (FEMA) recognizes a fallout threat. This corresponds to particle sizes that range from 50 μm to several millimeters in diameter.[3]

Although the amount of radioactive material is minute in comparison to the total amount of debris, the radiation levels are quite high. About 57 g of

Table II. Typical protection factor ranges relative to blast protection codes.[a]

Blast preference	Description[b]	INR	Fallout
A	Subway stations, tunnels, mines, and caves with large volume relative to entrances	10–10 000	1000–10,000
B	Basements and sub-basements of massive (monumental) masonry structures	10–100 100–1000 (sub)	100–1000
C	Basements and sub-basements of large, fully engineered structures having any floor system over the basement other than wood, concrete flat plate, or or band beam support	10–100 100–1000 (sub)	100–1000
D	Basements of wood frame and brick veneer structures including residences	4–8 (wood) 5–10 (brick)	10–50
E	First three stories of buildings with "strong" walls, less than ten aboveground stories, and less than 50% apertures	2–5	20–80
F	Fourth through ninth stories of buildings with "strong" walls, less than ten aboveground stories, and less than 50% apertures	1–5	20–100
G	Basements and sub-basements of buildings with a flat plate or band beam supported floor system over the basement	10–100	100–200
H	First three stories of buildings with "strong" walls, less than ten aboveground stories, and greater than 50% apertures; or first three stories of buildings with "weak" walls and less than ten aboveground stories	2–5	20–80
I	All aboveground stories of buildings having ten or more stories. Fourth through ninth stories of buildings having "weak" walls	1–5	20–100

[a] Adapted from Ref. 2, Panel 9 and Ref. 3, Panel 19.
[b] Load bearing walls are considered as "weak" walls.

fission products are formed for each kiloton (or 57 kg/Mt) of fission energy yield. At one minute after a nuclear explosion, when the residual nuclear radiation has been defined as beginning, the gamma activity of the 57 g of fission products from a 1 kt fission explosion is comparable to that of about 2.7×10^7 kg (30 000 tons) of radium in equilibrium with its decay products.[4] Therefore it is the early fallout that produces the most immediate health hazard.

Protection factors

For a sheltered population the greatest hazard is gamma radiation. This is because alpha and beta particles will not penetrate a shelter and neutrons are a problem for the initial nuclear radiation. Fallout gamma rays undergo the

Figure 1. Variance of protection factor inside a building (from Ref. 3, Panel 18).

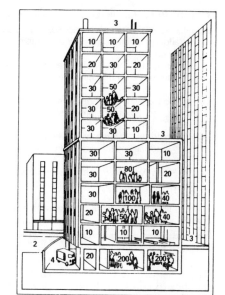

same interactions with matter as the initial gamma rays, but on the average are less energetic. Thus, in comparison with initial nuclear radiation, fallout is easier to protect against as can be seen from a comparison of the last two columns of Table II. Again, the lower value for the protection factor in the ranges of those quoted corresponds to openings in the structure.

The fallout protection factor, as given in Table II, is the usual protection factor (PF) that one finds in the literature. It is defined as the ratio of gamma radiation exposure at a standard unprotected location to exposure at a protected location. The standard unprotected location is defined as a point 3 feet above an infinite, smooth plane uniformly covered with fallout. Protection factor calculations also assume that fallout is spread uniformly on ground and roof surfaces. The shielding effect of nearby buildings is taken into account, but the movement of fallout by wind and rain is not.

The protection factor is not to be confused with the outside/inside ratio. The protection factor is a theoretical calculation whereas the outside/inside ratio is the actually measured ratio of the fallout gamma exposure rate at some point outside a shelter to the exposure rate at some point inside the shelter.

The protection factors in various parts of buildings are calculated from data obtained in the National Fallout Shelter Survey as conducted by FEMA. These data include the size of the building, construction material, building openings, and what type of buildings or open areas surround the shelter. The calculations use relationships derived from radiation penetration theory that describe how gamma radiation from fallout-contaminated surfaces is reduced in intensity as it passses through walls, floors, and air.

Figure 1 shows how the protection factor can vary inside a building. It should be noted that a protection factor is ascribed to open areas outside the building. This is because the protection factor is the ratio of dose rates with

Table III. Protection in residential basements.[a]

The fallout protection afforded by home basements can be estimated in the following way:

(1) Single-story homes with average basement wall exposures (i.e., aboveground) less than 2 feet will provide at least PF 20 throughout the basement.

(2) Homes with 2 or more stories and 2 feet or less average basement wall exposure will provide at least PF 40 throughout the basement.

(3) Single-story homes with average basement wall exposure greater than 2 feet can be improved to PF 20 by sandbagging the exposed walls or mounding earth against them.

(4) Similarly, multistory homes with basement wall exposure greater than 2 feet can be improved to PF 40 by sandbagging or mounding earth.

Generally, fallout protection in home basements is least in the center of the basement and greatest in the corners along the walls.

[a] Adapted from Ref. 3, Panel 21.

respect to the standard surface. Therefore pavement will give a PF of about 1.25; sand, bare soil, and grassy areas a PF of approximately 1.5; gravel roads a PF of 2; and very rough or plowed ground a PF of 2.5[3]

A PF of 40 has been determined to be the minimum desirable protection factor when looking at the U.S. as a whole. Frame homes provide a PF of 1.5–3.0 while basements have a PF of 10–20. Table III shows the estimated protection factors of home basements. As can be seen, many basements would need to be upgraded to offer the minimum PF 40. In addition to mounding earth around exposed basement walls, Fig. 2 shows a suggested plan for interior alterations.

Instrumentation

Because radiation levels can vary widely in a shelter, some means of measurement is necessary to indicate those areas that offer the greatest protection. Between 1962 and 1964 the Federal government purchased instruments to be placed in congregate fallout shelters. These instruments included 2.8 million dosimeters, which measure total radiation exposure, along with 300 000 "low" range and 600 000 "high" range rate meters which measure radiation intensity. No further instruments have been purchased for shelter use since the initial acquisition. Currently these instruments are on a four-year cycle for calibration and servicing.

The low-range meter (CD V-700) has three ranges: 0–0.5, 0–5, and 0–50 milliroentgens per hour. The high-range meter (CD V-715) has four ranges: 0–0.5, 0–5, 0–50, and 0–500 roentgens per hour. The CD V-700 is intended primarily for training and in the months after the attack for cleanup operations after the radiation levels have subsided. The dosimeters and the CD V-715 measure gamma radiation only, while the CD V-700 measures gamma intensity and has the capability of detecting beta radiation. The CD V-700 utilizes a Geiger-Muller tube as the detection device while the CD V-715 makes use of

Nuclear radiation and fallout

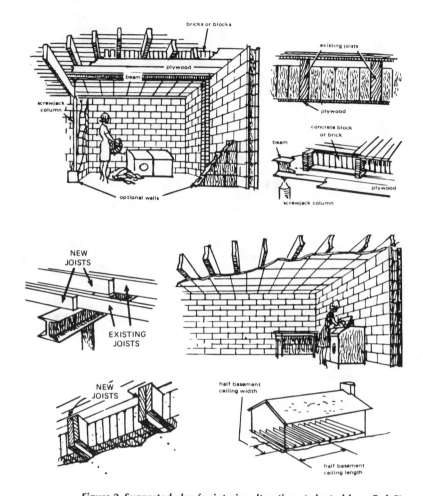

Figure 2. Suggested plan for interior alterations (adapted from Ref. 5).

an ionization chamber. Both meters are battery operated with a continuous operation battery life of 100 hours for the CD V-700 and 150 hours for the CD V-715. Intermittent usage of the meters can prolong the battery life dramatically.

Three different dosimeters are available for shelter use. Each has a different range: 0–20, 0–100, and 0–200 roentgens. Dosimeters require an auxiliary device to readjust them to zero. This device is the dosimeter charger and 400 000 of these were initially purchased.

All of these instruments are designated for congregate shelters only. People opting to remain in their individual shelter could purchase their own instrumentation if they so desired. Without detectors, people will have little knowledge of the levels of radiation they have been subjected to.

Civil defense

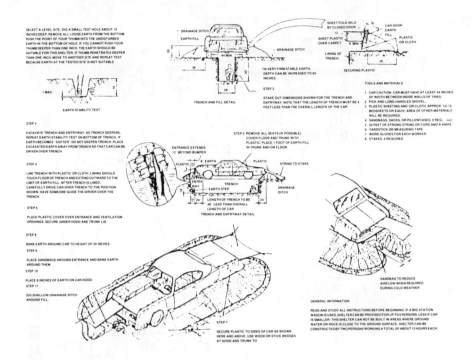

Figure 3. Expedient fallout shelter (car over trench).

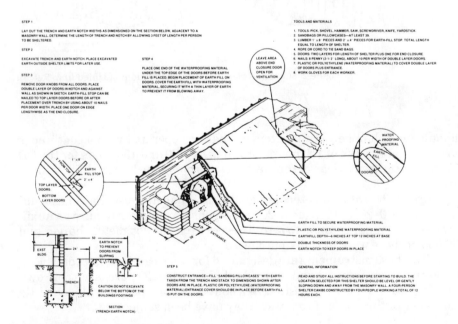

Figure 4. Expedient fallout shelter (tilt-up doors and earth).

Conclusions

People can be protected from nuclear radiation. The problems come mainly from economics and political will. Neither the government nor the people are willing to expend the amount of funds required to construct or identify either congregate or personal shelters. Because of the cost factor the Federal Emergency Management Agency has developed plans for expedient shelters that are claimed to offer a minimum PF of 40. Two of the more interesting plans are shown in Figs. (3) and (4). These FEMA designs for expedient shelters have provoked reactions ranging from ridicule to considerable interest depending on the individual's belief in civil defense, home area, soil conditions, experience with excavation and construction, and his or her physical condition.

Bibliography

The Final Epidemic, edited by Ruth Adams and Susan Cullen (University of Chicago, Chicago, 1981).
Last Aid: The Medical Dimensions of Nuclear War, edited by E. Chivian, S. Chivian, R. J. Lifton, and I. E. Mack (Freeman, New York, 1982).
The Counterfeit Ark, edited by Jennifer Leaning, and Langley Keyes (Ballinger, Cambridge, MA, 1984).
Nuclear Radiation in Warfare, by Joseph Rotblat, Stockholm International Peace Research Institute (Oelgeschlager, Gunn and Hain, Cambridge, MA, 1981).
How to Manage Congregate Lodging Facilities and Fallout Shelters, Federal Emergency Management Agency Report No. SM-11, 1981.
Radiation Safety in Shelters, Federal Emergency Management Agency Report No. CPG 2-6.4, 1983.

References

1. *The Effects of Nuclear Weapons*, 3rd ed., edited by S. Glasstone and P. J. Dolan (U.S. Department of Defense and Energy, U.S. GPO, Washington, DC, 1977).
2. *DCPA Attack Environment Manual,* Defense Civil Preparedness Agency Report No. CPG 2-1A5, 1973 (Chap. 5).
3. *DCPA Attack Environment Manual,* Defense Civil Preparedness Agency Report No. CPG 2-1A6, 1973 (Chap. 6).
4. *Radiological Defense Textbook*, Defense Civil Preparedness Agency Report No. SM-11.22-2, 1974.
5. *In Time of Emergency*, Defense Civil Preparedness Agency Report No. H-14, 1968.

Chapter 6
Sheltering from a nuclear attack

James W. Ring

Introduction

The effects of a nuclear explosion were discussed in Chapter 4 and are usually listed as blast, direct nuclear radiation, thermal radiation, fires, electromagnetic pulse, and fallout.[1] In a nuclear attack there may be many such explosions on different targets, and the effects of an attack will be both quantitatively and qualitatively different from those of an isolated, single explosion. Planning for sheltering should include ways of attenuating as many of these effects as possible. The methods which might be used to provide shelter from direct nuclear radiation and fallout are treated in Chap. 5.

One difficulty in making sheltering plans is that it is never easy, and often not feasible, to provide adequate shelter from all of these effects. Even the Swiss who are spending the most per capita on civil defense of any nation do not aim to protect their citizens from direct hits. Partly for this reason it is common practice to separate shelters into two classes, blast and fallout. The major problem near, but not at ground zero, which is blast, can be handled, with other problems such as direct radiation and fires dealt with as secondary effects. At a distance from ground zero where blast effects are not so serious the major problem is one of fallout. Different designs are called for to meet these quite different purposes and the expense of building such shelters is also markedly different. It should be pointed out here that the blast shelter is usually also a good fallout shelter although the converse is not true.

The distinction between blast and fallout shelters is a useful one and allows for a rational and practical approach to the sheltering problem. There are, however, still formidable difficulties which need to be surmounted. Many of these are related to uncertainties in the conditions which prevail before, during, and after the attack. Examples of some of these uncertainties are the warning time before an attack (is movement of urban populations possible?), the weather at and after the attack (what will the fallout pattern be like?), will large-scale fires be started? (will people in shelters still be safe?), will nuclear winter ensue? (what will the survivors of the nuclear attack have to face?).

This chapter attempts to deal with these questions by first separately describing the two kinds of shelters and discussing some of the uncertainties which in each case affect the efficacy of that kind of shelter. Next the problem of the attack strategy will be addressed and its relationship to the sheltering plans will be discussed.

Blast shelters

A 1 Mt explosion on the surface will produce a blast wave of at least 100 kPa (15 psi or about 1 atm) overpressure out to a radius of 2.4 km from ground zero while an air burst could increase this radius to about 2.7 km. The direct ionizing radiation effects for large weapons will be less in range than the blast effects and thermal radiation effects. The thermal radiation effects can be serious (third-degree burns) up to a radius of 8 km but if people are in blast shelters they will be protected from these effects except very near ground zero where even hardened shelters could be destroyed and radiation both thermal and nuclear would be so intense as to kill even if blast does not.[2] It should be noted here, however, that air bursts which devastate larger areas for equal yield do not produce as great overpressures near ground zero. Thus if blast shelters hardened to greater than 40 psi overpressure are provided *and* if the weapon is air burst, then even at ground zero people so sheltered could survive.

Centers of commerce, industry, transportation, government, and communication are often considered strategic targets and thus for the urban 70% of the population one might well consider providing blast shelters.

For the purpose of understanding the important elements of a blast shelter consider an example for which the Federal Emergency Management Agency (FEMA) has provided plans.[3] These plans give drawings and specifications for building a shelter of reinforced concrete providing for protection up to 15 psi overpressure for up to six adults. Ventilation is designed to be closed off when the blast wave is anticipated but can be uncapped afterward and fresh air provided by a hand-operated centrifugal blower. Fallout can be kept out by means of an air intake hood which can be stored in the shelter (FEMA calls it optional). The entry to the shelter is via a hatchway which is closed by a heavy (600 lb) wooden hatch cover and a smaller hatch door (125 lb) also of wood. Instructions are given for operating the blower, e.g., for 15 minutes every two hours, and for removing the caps, e.g., do not do so if smoke contaminates the air around the intake. The blast wave, if it enters the shelter through openings, can cause injury or death by acting directly on lungs or eardrums, by hurling debris at high speeds or by throwing people against other solid objects. Thus even the air intake and outlet pipes should be sealed when a blast wave might be passing.

Other larger shelters have also been planned (e.g., in Switzerland) and existing structures such as subways have been suggested as a means by which large numbers of urbanites could be sheltered. FEMA has also studied ways in which new buildings could be designed so as to provide both additional blast and fire protection.[4]

Some of the problems arising from the uncertainties suggested in the Introduction have already been suggested by the description of the family blast shelter given above. The necessity for sealing off openings both to prevent entry of the blast wave and to keep out smoke and toxic gases or fumes means that subways would have to be provided with blast doors and a new ventilation system to meet these specifications.

Other problems have to do with the other effects and their consequences on the people in the shelter. Fires above the shelter could mean not only smoke, toxic fumes, and CO, but also deprivation of oxygen if a fire storm or conflagration began in the blast rubble above the shelter. The temperatures associated with large fires feeding on this debris could get high enough to cause deaths in the shelter.[5] This does not mean that all or even most of those sheltering would be killed. For example, in the Hamburg fire storm a large majority of those inhabitants in underground shelters survived the fire.[6]

Fallout may also be a problem. The protection factor to fallout radiation offered by the FEMA shelter is much greater than 40, so the occupants are subjected to less than one-fortieth of the radioactive dose that they would experience outside. Fallout, however, can not be allowed to enter the shelter. If it does, the shelter must be decontaminated, and that may not be possible during or for some time after an attack. Even if the ventilation system is properly designed and operated so that contamination does not occur, there will be problems when people enter or leave the shelter. Fallout near ground zero can begin within 10 minutes after detonation from the stem of the mushroom cloud, and this early effect during the first hour after detonation represents a prime threat to emergency crews seeking to put out fires, shift debris, collect supplies, locate family members, or aid the wounded. The patterns of this fallout, like those of the later long-term fallout, depend upon wind and weather conditions and on whether the bomb exploded in the air or on the ground.[1] In the highest radiation areas (covering only a small fraction of the U.S.) the accumulated fallout can become high enough on the surface above the shelter so that even with high protection factors sickness and death may ensue within. Since high radiation levels may continue for some considerable time outside the shelter, provision for storage of food, water, clothing, and medicine should be made within the shelter, and sanitary facilities should also be provided. Radiation survey meters are essential to test the outside environment and to test for contamination inside the shelter.

Providing blast shelters is an expensive business. The Swiss who have embarked on a program to provide such shelter for every citizen have found this to be the case. The figures range from several hundred dollars (1978) to well over a $1000 (1983) per person. An American estimate reported by the Federation of American Scientists in 1981 is $1000 per person.[7] There are at least 100 million Americans who live in possibly targeted cities. If we followed the Swiss model we would provide blast shelters at a cost on the order of $100 billion. The Swiss have spread construction over decades to accommodate this cost (see Chap. 8).

Fallout shelters

For those who do not live in urban targeted areas or near military targets, e.g., SAC bases, or industrial targets, e.g., oil refineries, or for those urbanites who survive the direct immediate effects there is still the problem of fallout to consider. Idealized fallout patterns from ground bursts are given in Refs. 1, 2, and 7. For a 1 Mt ground burst a seven-day accumulated pattern resulting from a constant 15 mph wind would yield a 450 rem (roentgen equivalent mammal) dose contour in an oblong pattern 256 km long and about 64 km wide at 96 km downwind. This 450 rem dose is taken to be a lethal dose in 50% of the cases and hence in the roughly 10 240 square kilometers covered all people must evacuate, take shelter, or suffer death or at least severe radiation sickness. Much higher doses can accumulate over a more restricted area nearer the detonation point. Doses of more than 3000 rems, for example, can occur over an oblong of perhaps 48 km in length depending on the wind. For a large-scale ground-burst attack the areas targeted including all large cities would have two week accumulated doses of over 10 000 rems covering the cities and extending downwind perhaps many tens of kilometers.

Fortunately fallout shelters can be easily constructed in existing basements. With 46 cm of dirt or 30 cm of concrete intervening a protection factor (PF) of 10 can be achieved against fallout radiation. Thus with dirt piled up against the above ground outside basement walls, windows and doors closed and sealed in the house above, the door to the basement closed and sealed and an exhaust air duct provided, the persons in this shelter could take the 450 rem dose delivered outside and find the dose inside to be 45 rems or less.[8,9] However, even for a shelter with a PF of 40, which meets FEMA's standard, a dose of 10 000 rems outside would mean 250 rems inside and radiation sickness would ensue. For more detailed descriptions of fallout shelters see Chap. 5.

The uncertainties in the fallout pattern depend primarily on wind and weather conditions, but the local topography, the existence of fires (particularly fire storms or conflagrations), as well as other conditions can also affect the pattern.[1] Thus it is important for those sheltering from fallout to have means themselves to measure radiation levels. They must also be very careful about contamination of the shelter space and monitoring checks with survey meters inside and outside the shelter and of those people entering the shelter should be routinely carried out.

The expense of preparing such shelters is a great deal less than that for blast shelters.[9] Indeed, some have argued that many people who survive the initial effects would have sufficient time to pile up dirt around their basements, move supplies down there, and make other preparations before the fallout begins to arrive or at least before it accumulates to a dangerous level.[10] Survey meters and medical supplies are very important and would probably have to be provided beforehand. Also since a stay of up to several weeks may be necessary, stocks of food, water, and clothing should be provided and sanitary facilities arranged. These supplies and arrangements would cost something but the level would be a lot less than $1000/person and could be arranged by individuals without the need for government actions.

The importance of having some warning time is apparent especially if no preparations for fallout sheltering have been made in advance. With a day, or even several hours, a great deal can be done to make a basement an effective temporary fallout shelter.

The attack strategy

The Office of Technology Assessment (OTA) in its study, *The Effects of Nuclear War*, looks at several different kinds of attacks on the U.S. and the U.S.S.R.: an attack on a single city, an attack on oil refineries limited to 10 missiles, a counterforce attack, and an attack on a range of military and economic targets using a large fraction of the existing arsenals. Although these attacks are chosen somewhat arbitrarily, they are meant to represent a full range of possibilities and to include the more important possible strategies. Those obviously included are a strike at the nation's industry (oil refineries), a strike at the opponent's counterforce capabilities, a strike at the opponent's centers of transportation, industry, commerce and communication (a large attack on many facets of the economy) combined with an attack on military targets. In each of these cases millions would be killed with numbers increasing as the number of weapons increases. Thus in the final case studied there are estimated to be between 20 and 160 million immediate deaths with more tens of millions likely to die later either because the economy is unable to support them or because, in the longer term, cancer and to a lesser extent genetic damage show up and cause deaths. These and other long-term effects are treated in Chap. 9.

Another attack strategy which is implied by emphasizing civil defense as a strategic measure is an attack which seeks to nullify the advantages for recovery which a nation may gain by a highly organized, necessarily well known, civil defense plan which would include as a central part an overall sheltering plan for its population. If in the aftermath of a nuclear war that nation which can recover first is to be the winner then the opponent will wish to deny that possibility. This kind of strike might be termed a counter-recovery attack.

Another strategy which is not mentioned by the OTA, and is not the prime purpose of the counter-recovery attack, is that of mutual assured destruction. One interpretation of this strategy is that the enemy's population is to be held in hostage in the sense that it will be the target of the counter attack if a nuclear war is started.

The point of listing these possibilities here is not to argue for any of them as being more probable than others, but simply to emphasize the uncertainties that must be faced in attempting to make civil defense sheltering plans. The one conclusion upon which virtually everyone agrees is that any nuclear war will result in millions of deaths, and this will be the case whether there is a national sheltering plan or not. The difficulty, of course, is that we do not know the enemy's strategy. We do not know his targets and we do not know what changes he may make in his strategy and therefore his target choices as conditions change in the world, in his own country, or as a result of the war in which he is engaged.

To illustrate how difficult it is to make sheltering plans consider the following. A sheltering plan like that of the Swiss, who obviously do not expect to play the same role in a nuclear war as the U.S. or the U.S.S.R., would provide for U.S. citizens (or U.S.S.R. citizens) blast protection up to at least 15 psi overpressure and food, water, clothing, medicine, etc. for an extended stay in the shelter. The advantage of such a plan is that it does not involve mass movements of population, but rather on short notice (not days but minutes) urban populations could take shelter in places already prepared in deep basements, subways, tunnels, etc. The flaw in this plan is that once again the enemy need only direct some of its many weapons at these well-known targets and the strategic advantage of sheltering is reduced. The cost of such shelters is high, something like $1000/person, and if the metropolitan areas with populations of 200 000 or more are used as an example there would be about 130 million in the U.S. needing shelter or $130 000 000 000 spent in providing such shelter. The countermove by the enemy would obviously be much less costly to him as he is already equipped with 9 or 10 thousand such weapons and there are only about 150 such cities. Ten weapons for each city would thus consume only 1500 or about 15% of his inventory.

Yet another major difficulty in planning for civil defense comes not necessarily from a given but unknown strategic plan of attack but rather from changes in plans which may occur as the war develops. Most of the scenarios envisioned by the OTA, by FEMA, and other planners involve only one wave of attack (which turns out to be devastating enough) but suppose another wave occurs. The emergency crews are out working to clean up and repair essential parts of the system like roads, power lines, sewers, and water lines. People will be emerging from shelters to go back to work and to start putting together again the highly structured, organized society which we previously had. All will be subjected to lethal effects and once again be forced to seek shelter but this time with much less hope of eventual recovery. The fallout which for the most part dies out quickly will be renewed and augmented and the stay in shelters will thus be extended but with stocked supplies already diminished. If one adds to this tale of woe the climate changes predicted by a serious nuclear winter, then the chances for survival become very grim.

To paint a very black and probably unrealistic picture, consider what would be the result if all of the 9 000 Russian warheads were to be targeted on the U.S. were centered on cities, but overlapping in such a way as to devastate the maximum area. Recall that 70% of our population lives in cities which occupy only 2.5% of our land area. Each megaton would devastate an area of say 3.2 km in radius (12 psi) or about 32 square miles in area. Altogether then something like 256 000 square kilometers of area in and around cities would be completely devastated and fallout sufficient to cover many times over the entire area of the country with lethal dose accumulations would occur. Of course, there would not be a uniform distribution of fallout, but in highly populated parts of the country with correspondingly more weapons detonated, this would mean that the survivors of the initial effects would be at even greater risk from the greater amount of fallout. And in this case the downwind suburbs would be subjected to doses of many thousands of rems. So even with a shelter whose PF is 40, as FEMA recommends, the dose inside could be high enough to cause sickness or death.

What then can be done?

The facts and arguments given above make it clear that any strategic civil defense plan or even any civil defense plan which could be interpreted by the opponent as part of a strategy could be easily countered by the opponent at relatively little expense by simply targeting his many redundant missiles and warheads accordingly. On the other hand, the expense to the nation who develops and implements a blast shelter program will be very high as we know from the Swiss experience. Such strategic sheltering plans do not seem to be cost effective if the enemy deliberately targets civilians.

In the case of a nuclear attack those caught near ground zero will have little or no chance of survival, but depending on the size of the attack many people, perhaps a large majority of the population, will have a better chance if they can shelter from fallout. What would be needed are places which have a large enough protection factor; e.g., a basement has a PF of 10 which can be easily increased to 20 or more, and with more planning and preparation to the PF of 40 recommended by FEMA. These shelters should be supplied with such items as food, water, medicine, sanitary facilities, and a radiation survey meter. Descriptions of fallout shelters including ways of increasing PF's and preparing such shelters without basements are given in Chap. 6.

A study of the feasibility of sheltering from fallout in a basement (or other space whose PF is 20) has been made using a mathematical model.[11] The results show that even for a large-scale attack of 6559 Mt such already existing shelters would be feasible for most of the country and would not require prohibitively long shelter times.

The kind of life which survivors emerging from these shelters might face is dealt with in Chap. 10. The concern here is sheltering from the nuclear attack itself. And with such shelters there would probably be a better chance for short-term survival.

Summary

The aim was to examine both blast shelters and fallout shelters as components in sheltering plans and in particular to look closely at (a) the uncertainties in the effects of nuclear explosions and their interactions and how these might affect the efficacy of the shelters and (b) the uncertainties in the attack plan of the enemy and how these might affect the efficacy of the shelters and the sheltering plans.

We have found that the difficulties in the former set of uncertainties are formidable. The price for solving some of these problems and then implementing the solutions is high, and even then we cannot be confident of success as uncertainties such as fire and weather cannot be completely planned for.

Unfortunately, there is not even a partial solution to the second set of problems. We can never depend on knowing what the enemy's plans for attack are, and his attack plan is crucial in determining the outcome of our sheltering plans. Furthermore, the cost of building better shelters, blast rather than fallout or hardening blast shelters, is greater than the cost to the enemy of utilizing more missiles and warheads. Blast shelters thus do not seem to be a good investment.

Fallout shelters, however, seem the only cost-effective response to the problem. They may save many lives and they can be prepared by individuals, families or groups of citizens. What is needed is an educational program which would instruct citizens in how to construct fallout shelters and how to use them.

We all should understand that fallout shelters are not at all effective close to ground zero. They will not protect against blast, nor are they protection against large-scale fires. What they will do is protect against fallout whose dangerous range is much greater than any of the other effects except that of nuclear winter. Thus for those outside the blast and fire zones a better chance of survival could be achieved by sheltering in a suitable place. It should also be clear, however, that the more warheads detonated the less the chances for survival, even when fallout shelters are available. In an accidental or limited attack these shelters can greatly enhance the survival odds and even in an all-out attack large segments of the population can expect that fallout shelters will make it possible for them to survive.

References

1. *The Effects of Nuclear Weapons*, 3rd ed., edited by S. Glasstone and P. J. Dolan (U.S. Departments of Defense and Energy, U.S. GPO, Washington, DC, 1977). This is the standard reference for the effects of nuclear weapons. Chapter V is on damage from air blast, Chap. VII is on thermal radiation, and Chap. IX is on fallout.
2. Office of Technology Assessment, *The Effects of Nuclear War* (Allanheld, Osmun and Co., Totawa, NJ, 1980), issued by the Office of Technology Assessment, Report No. OTA-NS-89, 1979. This book supplements and complements Ref. 1 in that it deals with the effects of war rather than the effects of individual weapons. Several different scenarios are explored and estimates are given of casualties, deaths, damage, and chances of recovery.
3. *Home Blast Shelter,* FEMA Report No. H-12-3, 1983 (U.S. GPO, Washington, DC, 1983).
4. *Increasing Blast and Fire Resistance in Buildings,* FEMA Report No. TR-62, 1982 (U.S. GPO, Washington, DC, 1982); *Protective Construction,* FEMA Report No. TR-39, 1983 (U.S. GPO, Washington, DC, 1983).
5. *FY 1971–1975 Research Summaries*, FEMA Report No. RR-3 (U.S. GPO, Washington, DC, 1983); *FY 1976–1980 Research Summaries*, FEMA Report No. RR-4, 1983 (U.S. GPO, Washington, DC, 1983).
6. K. Earp, *Deaths from Fire in Large Scale Attack—With Special Reference to the Hamburg Firestorm* (British Home Office, Scientific Adviser's Branch, London, 1953).
7. *Effects of Nuclear War*, Journal of the Federation of American Scientists, **34** (No. 2) (1981).
8. *Home Shelter*, FEMA Report No. H-12-1, 1980 (U.S. GPO, Washington, DC, 1980).
9. *Shelters in New Homes*, FEMA Report No. TR-60, 1969 (U.S. GPO, Washington, DC, 1969).
10. R. Scheer, *With Enough Shovels* (Random House, New York, 1982). In Chap. 2 on p. 18 an interview with T. K. Jones is quoted in which he makes the statement, "If there are enough shovels to go around, everybody's going to make it."
11. R. Ehrlich and J. Ring, "Fallout Sheltering: Is it Feasible?" (accepted for publication in *Health Physics*).

Chapter 7
Maintaining perceptions: Crisis relocation in the planning of nuclear war

John Hassard

Until very recently, the Reagan and preceding administrations had planned to evacuate people from designated "high-risk areas" to "host areas" in the event of increased tensions in a confrontation with the Soviet Union. This was seen as both a deterrent to attack and as a means of ensuring casualties would be minimized should that deterrent fail. This plan has fallen victim to recent budget cuts. In this chapter, I summarize those plans and the opposition to them in their historical context, drawing on examples from plans proposed for New York State and New York City. In my conclusion, I discuss possible future evacuation plans in the implementation of any strategic defense initiative.

The evolution of crisis relocation

Plans for protection of the U.S. population have undergone many changes over the years. In the Kennedy era, largely following recommendations of a committee directed by Wigner and Taylor at Woods Hole the emphasis was firmly on the protection of people through the construction of blast shelters. Indeed, this committee's findings formalized a civil defense response that had been favored ever since the Soviet Union first exploded a nuclear bomb in August 1949. Blast shelters had been proposed by many agencies and the evacuation of cities was considered futile.

According to a report issued by the New York State Civil Defense Commission, "Evacuation of people from target areas to areas of assumed lesser danger...is still given attention, [but] it seems of questionable value with the reduced warning times now in prospect... Evacuation of people from antici-

Ed. Note: This chapter was written before the Chernobyl disaster in the Ukraine, which provided an example of a large-scale evacuation of a population at risk. Approximately 100 000 people were evacuated after a 36-hour delay over a period of more than a week. An exclusion zone of 30 km was established. This endeavor, which will no doubt reduce the long-term effects of this tragedy, was about 1500 times smaller than the full nuclear war crisis relocation plans described in this chapter.

pated heavy fallout areas...seems extremely difficult to organize and administer in view of the uncertainties of attack and the obstacles to accurate forecasting of fallout patterns."[2]

By the early 1980s, however, plans to evacuate some American towns at risk were highly evolved, and were often regarded as prototypes for other risk areas. It is instructive to examine how the viewpoint of whether or not to relocate changed, and how evacuation plans evolved to be merely part of a general civil defense mechanism known as the Integrated Emergency Management System (IEMS). This system regarded all emergencies as having certain common elements and sought to make the responses to disasters more efficient by planning in a fashion that addressed more than one disaster type. This integrated approach, of which more later, has been fiercely criticized within and without the civil defense establishment.

The Federal Emergency Management Agency (FEMA), the body which oversees planning for emergencies (both nuclear and non-nuclear), aimed to coordinate the individual states' implementation of the integrated plans over a seven-year period at the total cost of about 4.2 billion dollars. As can be seen in Chap. 3 of this study, nothing approaching this sum ever materialized. The merging of nuclear and non-nuclear emergency procedures in FEMA's IEMS plan really became important in the early 1980s. The concept of "all hazards, full spectrum" planning had, in fact, been introduced in 1972 at the newly formed Defense Civil Preparedness Agency (DCPA), but had failed to gather much momentum owing to budget strictures and the difficulty of implementing such plans. In 1975, Secretary of Defense Schlesinger directed the DCPA to start developing crisis relocation plans (CRP), but again, budget cuts made progress difficult. By the time of President Carter's Presidential Directive PD 41, of September 1978,[3] which marked an enhancement of civil defense and its identification as part of the strategic deterrent, the "dual use" concept was firmly entrenched in FEMA's thinking. It was claimed that the emergency personnel and methods used were similar in the case of both nuclear and nonnuclear disasters. It was proposed that coupling the two made logistic and administrative sense. This was formalized by Rep. Donald Mitchell's (R-NY) introduction of an amendment to the Defense Authorization bill for the 1981 financial year which became Title V of the 1950 Federal Civil Defense Act. The following year, this "dual-use" concept was further cemented by Title VIII of the 1950 Act. Not surprisingly, funding for nuclear war planning was deemed to be more easily obtained when it was attached to uncontroversial contingencies. In July of 1982, an amendment introduced by Rep. Edward Markey (D–MA) which aimed to halve the civil defense budget to about $150 million was defeated 240-163. And in May 1984, Rep. Richard Ottinger (D–NY) tried to ensure that no civil defense funds could be used to prepare for nuclear war, but this amendment was also easily defeated.

In March 1983, President Reagan proposed that U.S. scientists should develop a defensive system capable of eliminating the threat from ballistic missiles. Crisis relocation could be seen as an integral part of this plan, particularly among those analysts who, though supporting the President's viewpoints on nuclear weaponry, realized that a 100% effective "astrodome" defense was unattainable. Ironically, at a time that the Strategic Defense

Initiative (SDI) is gearing up to spend large sums, FEMA has been ravaged by huge budget cuts. The crisis relocation plans have apparently been an early victim, and as of March 1985, no longer figure in strategic plans. FEMA spokespeople have been rather reticent to discuss the reason it had been dropped, but *The New York Times* of March 4, 1985 reported[4] that a FEMA spokesman, Russell Clanahan, asserted it was not budget cuts but changes in the agency's plans that forced its rejection. This may all be moot, since by this time, crisis relocation had been rejected by about 120 jurisdictions governing 90 million people. It is, nonetheless, a fact that FEMA had planned to ask for $345 million for the 1986 fiscal year but had been allowed only $119 million by the Office of Management and Budget. Its 1985 budget was about $181 million. Only the funding for the protection of the leadership and continuity of government seems to have escaped: in fact it was increased from $140 million to $155 million.[5] To understand the reasons and implications of this major change in strategic thinking, I will examine crisis relocation as a form of population defense in the last 40 years.

The idea that blast shelters were the correct way to shelter people from bombs and missiles came easily to people whose previous major experiences had been in the World War II. Evacuation of populations at risk had been achieved with mixed success in Britain and Germany during the war, but most protection had been accomplished through shelters. In London, for example, there had been large-scale evacuations from 1939 to 1941. In fact, one and a half million people were evacuated in the days before the war started and large numbers of others went to the country spontaneously. After this period, relocations occurred in response to specific enemy actions. Most evacuees had returned by the time the Nazis started bombarding the capital with rockets. The incidence of V-1 cruise bombs peaked in July 1944, when their launching sites started to be overrun by the Allies. The V-2 missiles started hitting South East England in September of that year, and again official evacuations occurred. As before, however, large numbers of evacuees soon returned. Blast shelters, on the other hand, were used all through the war, and they worked well against the bombs then used. This experience was not lost on the civil defense planners in the U.S. in the years following the war.

With the Soviet development of the H-bomb, however, it was realized that fallout from ground bursts was more easily protected against than the direct blast and heat effects of the multi-megaton weapons that were possibly aimed at American cities. It was realized the private urban blast shelters would provide little protection against direct or nearby ground bursts unless they were extraordinarily deep and well equipped. Sheltering is dealt with more extensively in Chap. 6 of this study. The conclusion was that the rapid evacuation of urban high-risk centers was the strategy which had the most chance of saving the most people from the blast effects of nuclear weapons. Such an evacuation coupled to an extensive program of fallout protection in host areas would provide a comprehensive civil defense.

What crisis relocation entails

By 1982, the plan envisaged the evacuation of 150 million Americans in about 400 "high-risk areas" to about 2000 "host areas."[6] As the planners in New York realized in 1960, this might have to be accomplished in very little time. A report by the Stanford Research Institute for the Defense Civil Preparedness Agency—the Government agency that oversaw civil defense programs from 1972 to 1979— assumed as its baseline case that there would be three-days' warning before a nuclear attack.[7] This figure seems to be the one most often quoted for major urban centers. Other authors—for example, those responsible for a major study by the System Planning Corporation[8]—have required one to two weeks for this period (although one of this report's authors, Sullivan, has suggested "one to three days" in the *Journal of Civil Defense*.[9] Wigner suggested that there would be "days or months" though they recognized that the attack could come with little or no warning.[1]

To avoid the expected congestion, various schemes were incorporated into the evacuation plan. In Washington, DC and other cities, for example, people with odd numbers on their automobile plates would be expected to wait for their even-numbered neighbors to go first. It was anticipated that spontaneous evacuation would account for perhaps 10% of all total evacuees, perhaps lessening congestion still more.[10] In major urban centers, other means of transport would be utilized to the full. For example, in New York City, it was planned that 10.74% of the population at risk would exit by air, 13.52% by rail, 2.64% by water, 15.52% by bus, and 57.57% by private automobile.[7] Their destinations were typically 200 miles away, in upstate New York and in five Pennsylvania counties. Fully 20% of the evacuees would be required to travel over 250 miles, the usual fuel limit for automobiles, and some would drive over 400 miles.

Once the evacuees reached their destinations, there would be measures undertaken to protect them from fallout. FEMA aimed at identifying one and a half million fallout shelters for perhaps half of the evacuees. The remainder would construct what FEMA calls "expedient shelters." These consist, for example, of a trench with a car parked overhead and dirt piled up around it. The final component of the civil defense procedure was for the evacuees to stay in their shelter for at least two weeks. Radiological monitoring would, of course, allow this period to be shortened if possible. This phase of the operation was not dealt with in any of the plans that I studied, but an Oak Ridge National Laboratory paper prepared for FEMA proposed that the younger elements of the population would stay "in the shelter...or...in decontaminated areas for the first two years." People over a certain age, probably about 40 years, "will leave the shelters" to forage.[11] Since long-term cancers are the main threat from radiation at this stage, this strategy decreases the population's overall cancer rate, particularly for the child-bearing sectors of that population.

The fallout shelters were supposed to have a protection factor (PF) of 40; that is, harmful radiation would be attenuated by a factor of 40 by the shelter's massive sides and roof. It has been often claimed that this number is rather arbitrary, chosen because it is attainable; indeed, in the New York Governor's Report already quoted it was claimed that a PF of 100 was the absolute minimum necessary. For a shelter below the level of the earth's surface, which is large enough for four people, a PF of 40 typically requires over one ton of material to be positioned by each evacuee. There is sufficient uncertainty in the effects of radiation on people and in the individual circumstances in any given shelter that it is difficult to assess what PF would be adequate, or even how to define "adequate" in such unusual circumstances as nuclear war.

The trend towards "dual use"

As will be seen, nuclear war evacuation plans had their detractors from the onset. *The New York Times*, in a critical editorial on President Carter's plans, claimed that "the real concern of course, was not for the defense of the population against the Russians but the defense of an arms control treaty against the Senate."[12] The Federal Emergency Management Agency has had its plans ridiculed in the press and at public hearings across the country. As stated, the jurisdictions responsible for 90 million people had rejected crisis relocation, and it was obvious to followers of this plan that this number would increase.

It is important to assess to what extent changes were made and plans evolved in order to make the plans palatable as opposed to making them more effective in changing strategic circumstances. In other words, we should ask whether the relocation plans evolved to counter the threat of hostility from the American public or the Soviet war machine, and if it was the former, were their effectiveness against the latter compromised? Furthermore, were the disaster plans for the other emergencies which were lumped with nuclear evacuation plans degraded? Underlying all these questions is the most important of all, and that is whether having these plans would actually increase the security of Americans in the event of a nuclear war.

Since 1980, the crisis relocation parts of civil defense had been increasingly entwined—on paper at least—with evacuation plans for other disasters. The reasons given were simple: both sets often involved the evacuation of the population at risk and both were controlled by the same public officials. Both required rapid response to changing circumstances and therefore required highly coordinated and organized control. It had also been found that, by coupling the non-nuclear civil defense funding to the nuclear civil defense planning, the latter was more enthusiastically undertaken by local authorities.

We should attempt, therefore, to establish what similarities and dissimilarities exist between nuclear war and a non-nuclear war disaster. This enables us to assess how much this coupling is due to political expediency and how much is due to a genuine convergence of needs and actions. Unfortunately, we enter rapidly into the realm of scenarios, with their inevitable uncertainties. For the sake of example, however, we will consider the disaster planning for a nuclear reactor. This example has certain elements common with nuclear war. There are, among other factors, the likely presence of radioactivity, the fact that the cause is definitely man-made, and that there is widespread public mistrust and fear of the likely result.

Typically, in the computer-aided studies of reactor emergencies, evacuation plans fall into two categories: those where more than 12 hours warning is given and those rarer cases where less than 12 hours are available.[13] In the event of a disaster, most early casualties would occur within four miles of a reactor. Utilities and power authorities typically employ a ten-mile radius emergency planning zone (EPZ). Many different responses have been developed which are tailored to minimize deaths in any given eventuality. For example, in the event of the release of certain radionuclides, it is safer for the local population to stay indoors, even within the EPZ. In the event of a meltdown, it would be crucial for a population evacuation to occur, even far downwind of the EPZ. The direction of relocation should be perpendicular to the plume's major axis. Most expert witnesses suggest that a few tens of miles would be a sufficient distance to travel. Most importantly, plans and their instigators should be flexible and the actions to be taken easy to implement. Although the evacuation in a nuclear reactor disaster would appear to be many times easier than that for nuclear war, certain local authorities claim that they are unable to develop adequate plans.[14]

The extremely competent work done by FEMA and local civil defense authorities on a daily basis for chemical spills, flooding, and so forth throughout America demonstrates a great deal of experience with such planning. The setting up of an effective command structure is detailed in the evacuation plans for the event of nuclear war and the form of this hierarchy is not incompatible with that which operates in more conventional disasters. An increase in the efficiency of this structure through more sophisticated communications and improved infrastructures would be both the aim of nuclear war evacuation planners and a useful goal for conventional civilian defense experts. Current nuclear war evacuation plans, however, require whole districts to relocate along set routes which have been predetermined for that district. Furthermore, most conventional disaster plans rely on the rapid influx of help—in both materiel and manpower—from neighboring regions. So while the command structure could be similar, the actual responses to nuclear war and non-nuclear war situations are likely to be very different.

During appropriation hearings over the "dual-use" aspect of funding for nuclear war, claims have been repeatedly made that it was not possible to separate the components of nuclear and non-nuclear disasters. For example, in the 15 March 1983 session before the Subcommittee on HUD–Independent Agencies, Dave McLoughlin, the Deputy Associate Director for State and Local Programs and Support was asked how much of the 1984 request for the protection planning element was for non-nuclear planning. McLoughlin replied, "Again Mr. Chairman, this is the same dilemma we were faced with before. If our emphasis is on generic planning, which is where we are going, hurricane planning and nuclear attack planning, etc., it is hard to separate those figures."[15] This sort of statement makes the claim that it is not possible to separate items like protection against electromagnetic pulse, blast, crisis relocation of all risk areas, and so on, which are distinctly oriented toward nuclear wars, from items like contingency plans for chemical spills, hurricane alerts, partial evacuations, etc., which are clearly not. Since effective responses to "natural" disasters demonstrably exist even now, this "generic" assertion would imply that present-day civil defense systems are at least partially effective against nuclear war. This has never, to my knowledge, been claimed by CRP proponents. If it were, and it were possible to identify those extra areas which need work to make them more effective against nuclear war, one should easily be able to "separate those figures." In response to a question posed earlier, about whether existing plans have been degraded by being lumped in together with nuclear war plans, the answer seems to be a tentative "no." Those charged with operating such plans may disagree with me, particularly if they have seen scarce funding and other resources be diverted from non-nuclear to nuclear war plans. It is precisely their major differences which keeps the non-nuclear plans unviolated. Consequently, the credibility of "dual-use" proponents is somewhat diminished.

Would crisis relocation work?

In practice, of course, we have little data with which to test the theories of FEMA and its consultants. The most ambitious (and also the only complete) testing of nuclear war evacuation plans was done in 1982 during the so-called Ivy League exercise, in which those civilian and military leaders charged with leading America in the postapocalyptic era were shunted around, out of reach of hypothetical Soviet missiles, to command bases in Texas and Massachusetts. The success of this test, of course, tells us little about how well the rest of the population would have fared in basements and under their automobiles. One often hears comparisons made between the amount spent in nuclear-war planning per individual for these leaders and for the public. Using a 1982 figure from the Library of Congress we find about a dollar being spent on each member of the general population, but many thousand times that on each government official.[16] This sort of comparison, although appealing to nuclear weapons freeze advocates and newspaper editors, avoids the fact that the continuation of stable government after a nuclear war depends on the survival of society's leaders. A disparity exists, in other words, not because our society's leaders are necessarily more deserving of survival, but because of the assumption that the long-term survival of the masses depends on the short-term survival of the few. The masses therefore have a right and a duty to ensure that the few will be able to significantly enhance their survivability. If this criterion is not met, the rationale for the spending disparity is hard to justify.

Of the pertinent large-scale evacuations in America, we can consider those undertaken before and during hurricanes on the Gulf of Mexico and South Atlantic coasts (in particular, that for hurricane Carla) and that evacuation which occurred largely spontaneously around the Three Mile Island nuclear power plant in Pennsylvania. In 1961, for hurricane Carla, for example, between 1/2 and 3/4 of a million people were evacuated without a single major reported accident or fatality.[17] In the near meltdown at the Three Mile Island nuclear power plant, it has been estimated that 40% of the local residents evacuated spontaneously. This event has provided much information on the public's response to potential disasters.[18]

State civil defense authorities have had much experience and great success with small-scale evacuations caused by chemical spills and floods too numerous to mention. In all these cases, however, the differences with nuclear war are obvious and large. The crisis relocation plans called for the evacuation of numbers of people two orders of magnitude greater than for hurricane Carla. It would be naive to assume the problems would scale linearly. All major disasters to date have relied extensively on the rapid, ready, and willing help from outside the zone afflicted. This would not occur in a full-scale nuclear war. Soviet strategic doctrine repeatedly claims that their military would be unwilling or unable to limit a nuclear exchange. Since we cannot therefore assume that a nuclear war will be limited, extrapolation from these examples is not easy. While "stated Soviet strategic doctrine" and Soviet propaganda are often regarded to be the same thing, and that we therefore should not base our strategies on their stated intentions, the fact remains that there exist precious few moral or technical firebreaks in modern warfare, particularly with present-day accuracies and forward positioning. It would be foolhardy to assume that nuclear war will be containable, and to plan as though that were so.

The nitty-gritty of saving America

With so little experimental data with which to work, we are left with the plans themselves, or their prototypes, if we want to assess how effective they would have been in saving American lives. In common with another strategic system, the SDI, it has been difficult to obtain a definitive version of the crisis relocation plans which the proponents are prepared to defend. The Civil Defense Office of New York State at least had the courage of their convictions to commit to paper some prototypical plans and to offer them up for public criticism. Although it is true that criticisms of mere prototypes are not optimal since the real plans could change, it is possible to identify components of the plans that would inevitably be included in any arrangement to evacuate New York City, for example. Civil defense officials made it quite clear that the Plattsburgh and Utica/Rome plans were written in a style easily "exported" to other localities. Furthermore, New York State provides a balance of rural and urban communities which can be seen to have many features common to other states. It contains what must be the most difficult evacuation, that of New York City, but it also includes the town which is surely among the easiest, the town of Plattsburgh.

The plan taken to be prototypical in New York was that developed for Plattsburgh on the state's northeastern edge.[10] Approximately 44 000 residents were to have been evacuated along eight routes to areas between 15 and 30 miles from the city and its Strategic Air Command (SAC) base. Making reasonable assumptions about the number of people per car, and their driving speeds, one can calculate that, indeed, it would have been physically possible to remove that many people to their destinations in about 3 hours. Evacuees would be assigned to either private families in the host area or check in at one of the several public shelters. They would be required to park and leave their cars, hand over the keys, and, if necessary, perform various service tasks for their hosts. It is at this level of consideration that crisis relocation works best, since we do not, and cannot accurately, take into account breakdowns, stoppages, gridlocks, an unenthusiastic or scared populace, and people getting gasoline or lining up in the food stores. It ignores the fact that the only major route south passes right by the SAC base and another route out is a notorious one-lane bottleneck. Moreover, the extrapolation up from Plattsburgh, NY to any of the other ten identified risk areas in New York State is fraught with uncertainties.

Proponents of the plan conceded that difficulties remained, but pointed out that with more time and money much more planning could be done. Despite some famous and absurd statements[19] by some FEMA spokespeople, many supporters have produced cogent and sometimes passionate arguments in favor of nuclear-war evacuation. Their central thesis is as follows: since there is a nonzero probability for nuclear war to occur, we should seek ways to minimize its effects on society. If we assume that, for the foreseeable future, the U.S. will have no 100% effective defense against intercontinental ballistic missiles (ICBM's), we must conclude that nuclear weapons would be used either directly or indirectly (in counterforce attacks) on the U.S. population in the event of a war. Since there is, as yet, no practical way to shield a population from the direct effects of nuclear weapons (except by enormously expensive blast shelters), the only alternative is to remove that population from areas where they are most likely to be hit. Furthermore, since many people would spontaneously evacuate in an emergency likely to result in nuclear war, it makes sense and is morally correct to organize that evacuation so that it can be achieved on a more equitable basis. Finally, proponents note that, while there are assuredly logistic uncertainties in all such planning, the certainty of the alternative makes the choice whether to evacuate or not an easy one.

From talking with FEMA personnel, I have been left with no doubt that these planners were aware of many of the problems discussed here, but were convinced that given sufficient time and funding, the remaining logistic problems could be cleared up. There are many examples of areas where more work was needed. To be able to accurately assess crisis relocation's chances for success, it is instructive to look at the details of the plans. We can thereby examine what they require from, and assess the impact likely upon, the population at large. Lest you think that a civil defense planner's lot is an easy one, I will now list some, but by no means all, of the problems they face in crisis relocation.

Many states envisaged evacuating populations at risk across state lines. The example given of New York City residents heading for Pennsylvania is such a situation. FEMA recognized that a greater degree of interstate cooperation in planning would be needed. Even within a state, certain key personnel were clearly unaware of the extent of their roles. For example, in the evacuation plans of Plattsburgh, large numbers of city residents were due to travel to nearby Dannemora. In testimony to state legislators, the town supervisor of Dannemora claimed that he knew very little of such plans.[20] He added that, had there not been hearings, he would have known even less. While this is perhaps a trivial case, it both betrays the inadequacy of the plan and reveals the intricacies with which it must deal.

As will be obvious, the implementation of crisis relocation plans relied heavily on reliable communications systems. In the plans for Plattsburgh and Clinton County, NY, for example, medical resources and actions would have been through the standard telephone system with the sheriff's office in two-way radio contact with each medical station. In Utica and Rome, NY, public guidance would be accomplished with the help of local TV and radio stations, with, again, heavy reliance on the standard telephone network and "high power portable units" for emergency forces operating in the various zones.[21] The New York telephone company would have installed the necessary telephones in the Emergency Operating Center during the mobilization period (Ref. 21, Annex A, p. 23, undated) codenamed the "Sunlight" phase (Ref. 21, Annex A, p. 4, undated). One of the less well understood phenomena associated with nuclear detonations is that of the electromagnetic pulse (EMP). Any electrical equipment attached to an unshielded wire is vulnerable; the effect on communications during the attack (code named "Thunderbolt phase") could be catastrophic. In simulated wars, many hypothetical Soviet attacks are, in fact, initiated by several extremely high altitude nuclear detonations, or precursors.[22]

Another potentially disastrous aspect of crisis relocation concerns the health and medical care of the evacuees. One might imagine that the rapid evacuation of 150 million Americans and their subsequent sheltering in the host areas could cause problems in and of itself. In Plattsburgh, and in the prototypical plans for Utica/Rome, NY, evacuees were assigned either 10 or 20 square feet of living space for their two weeks in shelters. Given the stress of this situation, and the likely health consequences of living in such close-packed conditions, one could reasonably expect severe health problems among many of the evacuees even without a nuclear war. Again, studies of these conditions—overcrowding, the stress of an unprecedented disaster with no hope of outside relief—are (thankfully) restricted by the lack of experimental data.

The understanding of medical conditions in the event of nuclear war is also limited by lack of experience (and also probably our imaginations). Most studies have used data from Hiroshima and Nagasaki and from the Texas City disaster, a conventional explosion that killed over 4000 people. Abrams points out that 80% of the medical facilities in the U.S. exist in those cities with greater than 50 000 inhabitants, that is, the likely targets in a nuclear war.[23] The relocation of the urban population without its support systems could cause short-term inconvenience but long-term disaster.

The underlying hopes and assumptions

Throughout the plans and their prototypes we see many times the proviso that 100% public (and, of course, official) cooperation is assumed throughout. In Plattsburgh, however, 1700 critical workers would have been required to commute in and out of the risk areas to maintain essential services. In the Utica/Rome plans, the number was 43 000. Perhaps that is possible, but the plans also stipulated, for logistic reasons, that these workers families be relocated close to the risk areas.

One warhead per target was usually assumed. This ignored the doctrines of using at least two warheads per target, particularly in counterforce attacks. The plans ignored synergy between competing mechanisms of destruction. The option of retargetting missiles to strike at the relocated population, while of dubious military utility, is an option open to the enemy. If the enemy felt its deterrence capabilities being eroded by evacuation, it could exercise this option. Retargetting would make evacuation futile.

Simple topographic details were regularly ignored. For example, in the Utica/Rome plan, the authors defined the risk area west of the city of Rome by the Rome municipal boundary. The Oneida Lake plain is very flat from Rome all the way towards Syracuse, with no topographic feature (except, of course, distance) to reduce weapon effects. The Rome residents relocated there would be very vulnerable, but probably better off than their neighbors from Utica who are relocated east (that is, downwind) of Griffiss Strategic Air Command Base. Of course, the whole area is downwind of the Rochester, Syracuse, Buffalo, and Niagara risk areas. This was a problem with the plan, but not the fault of the planners. It is unavoidable in New York.

The most complex issues involved in the study of crisis relocation are those relating to the potential for human conflict. Except in the Southern states, the racial and social compositions of urban ("risk") populations are very different from those of suburban and rural ("host") classes. The possibility of conflict has long been recognized.[24] In most relocation plans one finds contingency appropriation plans for material and even labor needs. This is another area with potential for far-reaching civil unrest.

The economic effects of crisis relocation are similarly complex, and clearly correlated with the preceding problems. Laurino *et al.*, have made a comprehensive study and conclude that even with no attack, the economy could take at least one to two years to recover and the evacuation could cost about $100 billion.[25] Katz points to several assumptions made by Laurino *et al.* which suggest these conclusions are optimistic.[26] While this would indeed be a small price to pay if it saved the U.S. from an unavoidable and unique event, several studies have suggested that crisis relocation could become a regularly exercised strategic option.[27] The whole process would be undertaken several times each decade.

The strategic implications of crisis relocation

If it were only logistical problems associated with crisis relocation one might be tempted to conclude that the plan could, under certain circumstances, work, and in any case, we might as well have it. The human spirit is capable of extraordinary feats in times of need. As I have noted, there is a large uncertainty associated with the assumptions underlying crisis relocation planning. These and many other unknowns would conceivably have yielded to sufficient funding. However, the magnitude of that funding is not known. Furthermore, even a perfect understanding of a problem does not guarantee that a solution exists.

Unfortunately, it is not just the logistical uncertainties which formed the basis for much of the criticism of crisis relocation. The idea of moving 150 million Americans from their homes and places of work in a period of tension seems to many analysts to be so disruptive, chaotic, and provocative that it is unlikely that it would ever be ordered by an American President. Nonetheless, certain aspects have been singled out as being particularly improbable or dangerous or provocative.

It is this last aspect which deserves most consideration. While many authors have considered that the existence of fallout shelters are largely innocuous, the same studies have questioned the evacuation part upon which crisis relocation hinges. The Federation of American Scientists, for example, once thought that "the existence of a shelter program...could lower the provocation threshold, which might of itself make nuclear war more likely,"[28] but twenty years later saw "little strategic significance" in fallout shelters but pointed to the difficulty of predicting a war to take advantage of evacuation.[29] In order to assess crisis relocation's strategic implications, we will first consider the rationale behind CRP.

Crisis relocation's place in the strategic thinking of nuclear-war planners can be succinctly illustrated with a quote from Lt. General Daniel Graham, a former director of the Defense Intelligence Agency and founder of the "high frontier" missile defense advocacy group. "It is something that can happen today...The scenario starts when I receive intelligence reports that the Russians are evacuating their major cities. Soon a message from Moscow comes to the President [and]...points out that Soviet civilians have dispersed and in the event of nuclear war, Russian losses would be 10 million persons or less, compared with 100 million in the United States...It might add that it isn't a bluff because Russia lost 20 million people during World War II and 10 million isn't considered a big loss by the Soviets."[30]

This may seem to be an extremely hawkish view both of the Soviet Union and of international diplomacy, but it is a quote used extensively in FEMA publications. This Soviet casualty figure has also been suggested by other workers in this field. For example, it has been claimed that Soviet civil defense—consisting of blast shelters, evacuation and a week's notice—could reduce Soviet casualties to about 4%.[31] In other words, civil defense planning in general and crisis relocation planning in particular is regarded as useful to reduce the possibility of Soviet coercion.

Just as the nature of civil defense has changed in the last two decades, so have its objectives. In a message to Congress in 1961, President Kennedy asserted, "This administration has been looking hard at exactly what civil defense can and cannot do. It cannot be obtained cheaply. It cannot give assurance of blast protection that will be proof against surprise attack or guaranteed against obsolescence and destruction. And it cannot deter a nuclear attack."[32]

Kennedy then went on to justify his civil defense program as an "insurance...in the event of catastrophe." In President Carter's Directive No. 41, however, civil defense's role in deterrence was stressed[3] and since then, President Reagan has confirmed that "Civil defense, along with an effective continuity of government program, emergency mobilization and secure and reconstitutable telecommunications systems is an essential ingredient in our nuclear deterrent forces."[33]

Proponents of nuclear war civil defense planning rely heavily on the deterrence aspects of having a surviving population. Gouré, for example, points out that "(Soviet spokesmen contend)...that no country can rationally and credibly threaten nuclear war if it accepts that such a war would be suicidal. Thus, the credibility of deterrence in the nuclear age depends not only on a country's strategic offensive capability but also on its ability to convince itself, and especially its enemy, that it can survive a nuclear war, and therefore, that it can rationally threaten to resort to war if this proves necessary."[34]

This can be taken one small step further. If we can convince the enemy that we believe that we can survive (even though we know we cannot), our deterrence is made more credible. Analogously, the opponent's deterrence capabilities (or aggressive first-strike nuclear weaponry) become more threatening when we learn that, far from allowing his population to be held hostage by the specter of Mutual Assured Destruction, he is busily preparing to survive.

Jones elaborated this point at a Joint Committee on Defense Production: "There is a widespread belief that nuclear war would inevitably destroy both the United States and the Soviet Union...The avoidance of war does not necessarily depend on what Americans believe. It depends on what the leaders of the Soviet Union believe, even if their belief is ill-founded."[35]

The question may therefore resolve itself into several parts. Firstly, is the Soviet civil defense as effective as many fear? Would the fact that the Soviet Union was emptying Moscow frustrate the U.S. strategic targetters? This issue is addressed by many authors, including Harrell in Chap. 8 of this study. Secondly, can we convince the enemy that we believe our population is sufficiently invulnerable that our resolve to retaliate—if necessary with nuclear weapons—is unshaken by the prospect of global thermonuclear war?

In their new role as part of the strategic nuclear deterrent, many critics of the plans expressed disquiet at the implications of this policy. If, however, we neglect the moral controversies, we can take a more pragmatic approach and try to assess whether we are made safer by the existence of an evacuation plan before a nuclear attack and whether our long-term survival likelihood is increased after the attack.

For a deterrent to be effective, an enemy must perceive that he stands to lose more through undertaking a given course of action than if he restrained himself. In this context and in the wording of Presidential Directive 41, "Civil defense, as an element of the strategic balance, should assist in maintaining perceptions of that balance favorable to the U.S."[3] Consequently, if the Soviet Union does not believe that our civil defense through crisis relocation is effective, then it would not deter. Much evidence suggests that the Soviet equivalent is deemed largely ineffective by both East and West.[36] Naturally, there are exceptions to this viewpoint, and they are summarized in Chap. 8. Moreover, there are indications that the Soviet Union is shifting towards the protection of its urban population by means of blast shelters.[37] One oft-quoted Soviet expert, Leon Gouré, claims that 60% of the Soviet population is protected by blast shelters.[38]

We should also indicate the magnitude of the problem for the U.S. There are only about 545 cities in the U.S. with populations greater than 25 000. Many, probably a majority, of these were set to be "host areas" in the event of nuclear war evacuation. One can see that this is so by considering that there are only eleven designated "risk" areas in a state like New York. All other major towns are therefore "host" candidates. Ithaca, for example, where I live, has a population of about 29 000 (without students) and according to unsubstantiated rumor, was slated to receive 280 000 folks from Westchester County, NY. According to a report by the United States Arms Control and Disarmament Agency, a 200 kiloton warhead will produce effectively 100% casualties in a town of 50 000.[39] The Soviet Union has in excess of 6000 warheads of yield greater than this value. It is common knowledge—certainly to Soviet strategic planners—that ICBM's can be retargetted in a period of time far briefer than the shortest relocation period. On the evidence of the numbers of weapons involved, it would not appear that they would need to retarget if the destruction of American lives were their intention. Maybe they would spare Ithaca, but with the presence of a major university and the Seneca Army Depot up the lake, it is not obvious.

As regards retargetting, some CRP proponents contend that even if the Soviets were trying to maximize civilian deaths, evacuation would still result in relatively few casualties. For example, Sullivan et al. claim that, given an $11 billion five-year program, there would be 60% casualties in the event of a nuclear war with the Soviet Union.[8] If the population were relocated, there would be only about 10% casualties if the population were not targetted. If the Soviets chose to be as mean as possible, the casualty figure would rise to only 20%. Interestingly enough, this survival rate is nearly as high as that claimed for an approximately $70 billion shelter program studied in the same report.

The Soviet Union, however, need not rely on T. K. Jones or even the System Planning Corporation to determine how effective a nuclear attack on the United States would be. They have conducted many tests of their weapon systems, and presumably know their missiles' accuracies at least as well as does the CIA. They will be able to establish the likely implications of such an attack, and have repeatedly pointed out that they believe there would be no winners in a thermonuclear war, and certainly not the U.S. They are also aware of U.S. public sentiment and its abhorrence of the prospects of full-scale nuclear war—indeed some would say that the Kremlin has manipulated this sentiment. It seems unlikely, therefore, that they will ever really believe U.S. claims that Americans think they will easily survive such a war. The Soviets are also unlikely, therefore, to give much credibility to the posture of crisis relocation as a means of showing resolve.

If we are left with a strategy that does not convince our likely opponent, and therefore will not deter him, we would be prudent to examine whether having that strategy can actually harm us (beyond the normal problems associated with self-deception). Unfortunately, we again must resort to using largely untestable scenarios.

If the Soviet Union were to one day start emptying its cities, this crisis relocation could be interpreted as a sign of immutable resolve, but would also be seen by some as a prelude to a nuclear strike. Similarly, the Soviets would be correct to see the evacuation of the cities in the U.S. at the President's order as an escalation at the very least. As a response to a Soviet evacuation, the order to crisis relocate the "population at risk," while seemingly more justifiable on moral and strategic grounds, presents several problems. The process, although unlikely to be achieved in less than weeks, is relatively brief compared to the time it takes to make rational decisions on such vital matters. It is by no means certain that any given set of circumstances will have been foreseen and actions preplanned. Given the uncertainties and the stakes, decision making would be extremely difficult. The time constraints of modern warfare may precipitate rash or hurried judgments—no one would deny that crisis relocation is an escalation in an inherently unstable situation. One lesson of history relearned in every major crisis is that decision making and rational judgement tend to go astray on at least one of the protagonist's sides. Actions in such crises should therefore be made that stabilize rather than "up the ante." Perhaps crisis relocation, if it could somehow happen instantaneously and bilaterally, could stabilize such conditions under certain circumstances. This, of course, cannot happen.

The possession of a crisis relocation plan can be, and is meant to be, interpreted as an unwillingness to back down in such a confrontation. While this posture of firmness and resolve need not be provocative when confronted by a wily and ruthless enemy (and may indeed be quite the opposite), the means of demonstrating this resolve needs to be carefully examined.

Coupled with the highly accurate nature of many of the U.S.'s newest generation of nuclear weaponry—the land-based portion of which to a certain extent is itself vulnerable to the latest Soviet land-based missiles—such an act could be easily misconstrued as a prelude to an all-out attack. In other words, crisis relocation could actually precipitate the disaster it was seeking to mitigate. We have no way of knowing whether this would definitely happen, but we also have no way of proving that it would not. The political aspects of crisis relocation are dealt with more fully in Chap. 11 of this study.

We should now examine whether a relocated population would have a greater survival rate than one which had not evacuated. One viewpoint is that, since it cannot have a survival rate less than the alternative, it must be worth doing. This viewpoint, however, is invalid, if, for the reasons outlined above, by relocating ourselves, or by just having a plan to do so, we have increased the likelihood of the nuclear war. From a purely pragmatic point of view, it could be added here that a population in cars, trains, and buses is far more vulnerable to the direct effects of nuclear weapons than it would otherwise be.

One mechanism—hopefully not too likely—whereby merely having such a plan increases the risk of nuclear war, is if it made the President overconfident or overaggressive in his dealings with adversaries in the event of "increasing tensions." CRP proponents often draw analogies (see Appendix A.9) between having evacuation plans and wearing a seat belt in a car, or even more often to having lifeboats on a ship (this analogy lending itself nicely to visions of the Ship of State crossing the turbulent waters of the 1980's). They point out that having lifeboats does not make the captain take risks with icebergs. Opponents might find this analogy amusing, but probably also a trivialized simplification of rather complex issues. It contains the suggestion that nuclear confrontations are out of our control, like icebergs in the night. At the 1982 hearing in Plattsburgh on crisis relocation, a CRP proponent offered a slightly different analogy, likening the possession of crisis relocation plans to building houses with fire escapes. The chairman of the hearings, State Assemblyman Mark Alan Siegel (D-Manhattan) asked whether the CD official would use a fire escape if the whole street and town were on fire. Ship enthusiasts might ask themselves what would be the point of lifeboats that took "one to two weeks" to launch and then did not float. I would hope that they would do their best to warn other passengers, and to pressure the skipper to seek safe routes. They may even choose a new captain.

Life after the bomb

There has been much interest in the societal conditions in the years after the war, a great deal of it in Hollywood and the Hudson Institute,[27] but FEMA's crisis relocation plans have generally avoided the issue. The long-range recovery from nuclear war is discussed in Chap. 10 of this study. The lack of any real data has hampered this sort of work—few, if any, events in the human existence have much relevance. In her 1978 book, Tuchman studied the conditions in Europe in the fourteenth century following the Great Plague, in which at least a third of the population was wiped out.[40] As in nuclear war, the initial impact seems to have been in the denser concentrations of human habitation; the social fabric was largely shredded and disorder was widespread in the years following the disaster. Even more pessimistic is the conclusion in Ehrlich's "North America After the War," which suggests that a new ecology would take thousands of years to stabilize.[41]

In contrast to these accounts, a 1979 report from the International Center for Emergency Preparedness has summarized the research into these conditions by proponents of evacuation and admits to, among others, the following beliefs: "[In the post-nuclear war world...] (1) More than adequate food and water supplies would be available and the number of people who would die due to lack of food would likely be very small. (2) Large-scale disease or epidemics need not, and probably would not occur. (3) The degree of industrial damage that would be expected from a major attack would not produce insuperable bottlenecks. (4) Increases in the incidence of cancers in the surviving population would place an unimportant social, economic, and psychological burden on the surviving population. (5) Long-term ecological effects would not be severe enough to prohibit or seriously delay recovery. A nuclear attack could not induce gross changes in the balance of nature approaching the ones that human civilization has already produced (e.g., cutting forests, overgrazing hillsides, etc.).[42]

Despite this benign viewpoint, and its optimistic implications for civil defense, the details of the postapocalyptal world are largely unknown. It is not even clear that there would be a well-defined end to such a war. With the instigation of decapitation strategies and the deployment of easily hidden and dispersed cruise missiles by both the Soviet Union and the U.S., it is likely that this world could experience continuing nuclear warfare for long periods after the initial holocaust.

There is little agreement among specialists even on general aspects of life after nuclear war. Most people who would claim to be in the peace movement would agree that life would be nasty, brutish, and short. Others, for whom the term would be derogatory, but with access to the same facts and subject to the same physical laws, would vehemently disagree. Nonetheless, fundamental questions are still being posed. For example, the nuclear winter of Turco et al. not only questions the rosy post-war scenarios wherein U.S. society is rebuilt and prospering within five years but casts some doubt on the survival of the human species.[43] It was an encouraging sign that FEMA and the Pentagon decided to collaborate on a research study to check the nuclear winter findings and regrettable that this study also seems to have been a victim of budget cuts. The Pentagon, however, seems now to be accepting the major points made by Turco et al. and a study it commissioned calls for more research to be made in many critical areas.[44] President Reagan has even used nuclear winter as an argument in favor of his strategic defense initiative.

In contrast to this, there have already been refutations of the substance of the nuclear winter study. In particular, Edward Teller claims that the amount of water vapor in the atmosphere after a nuclear war would be a 10 000 times the amount of soot and, as the principle cleansing agent of the atmosphere, has not been correctly factored into the mechanisms of soot dispersal.[45] This in turn has been refuted and it is clear that the debate will continue.[46] However, this is the level of uncertainty which bedevils all crisis relocation plans. Ultimately, the most lasting impact of the discovery of nuclear winter may be an acceleration into research and development of very low yield (one or two kiloton) weapons, capable of earth penetration, and accurate enough to hit individually specified buildings. Nuclear winter is dealt with at greater length in Chap. 9 of this study.

Since we are dealing with the highest imaginable stakes, and with so many intangibles and unknowns, we should surely test any plan against very bad possible combinations of events—especially if they do not seem particularly improbable. In other words, crisis relocation plans—and in fact all civil defense planning—should be subject to the same philosophical requirements imposed on strategic planners: that the enemy is highly effective and our forces are highly fallible. However, this philosophy has not been evident.

Making plans for contingencies which are considerably less severe than those possible or likely, simply because it is possible to counter them, is acceptable (though regrettable) if one does not lose sight of their inadequacy. Incomplete or shoddily thought out plans or unworkable ideas should not be touted or sold to the public as effective: that is deceitful and can be counterproductive. The public should be aware of the danger it faces from nuclear war and that the only real solution to this problem is prevention.

Beyond the Astrodome

A recent development of extraordinary importance has been the apparent intention to research, and then presumably build, a system capable of "rendering [strategic nuclear missiles) impotent and obsolete": President Reagan's strategic defense initiative. The scale of such a system beggars the imagination—many estimates put the cost at about one trillion dollars and its implications are probably among the most controversial topics in the history of the arms race. One side sees this approach as the only way to remove the threat of mutually assured thermonuclear destruction; the other side finds it, in the words of Professor Hans Bethe, "difficult to imagine a system more likely to induce disaster."[46]

President Reagan's original idea was that the antimissile system would protect American (and European, and Soviet) cities in the so-called Astrodome defense. Although there has been no sign that the administration has lowered its sights, most workers in the field now talk increasingly of point missile defense to reduce the chance of an enemy first strike. If such a partial system were implemented, it is quite conceivable that the crisis relocation program could be reinstalled at a relatively small incremental cost. After all, 1% of a trillion dollars would buy the system that the System Planning Corporation claims would reduce casualties to 10% of the population. It would be rash, therefore, to claim that crisis relocation of high-risk areas is no longer on anybody's agenda. It may resurface when, as inevitably will happen, the SDI's planners realize (and can convince the administration) that defending all U.S. cities from nuclear devices is impossible.

Conclusions

I would conclude that crisis relocation as a separate entity or as a component of an integrated emergency management system is not an adequate response to nuclear war. I believe that the issue serves only to distract people from the reality of facing up to the only real solution—ensuring such a war never takes place. I see none but the most contrived and implausible reasons for supposing that such a plan could act as a deterrent, but have identified several ways it could act as a force of great social disruption and even precipitate the disaster it was devised to alleviate.

Whether this could change in the future is beyond this analysis, but the plan would, in my opinion, need to incorporate the following features: (1) Sufficient understanding and support of the plan by the U.S. Population that the effect of both active and passive opponents must be negligible; (2) A means of predicting more accurately the advent of nuclear war whether initiated by ICBM's, cruise missiles, depressed trajectory ballistic missiles, bombs in containers, on boats or in suitcases; (3) A means of protecting the population after they have arrived at their destinations, whether or not there is a nuclear war, and whether or not the disastrous post-war climatic predictions actually are correct; (4) The imposition of a strict, presumably military, public order system for the actual evacuation and the aftermath; (5) A means by which this system could be perpetuated during the possibly long periods of protracted warfare which may occur after the initial holocaust.

References

1. E. Wigner, *Project Harbor Summary Report,* NAS/NRC Publication No. 1337, Washington, DC, 1964.
2. Committee on Fallout Protection, New York State Civil Defense Commission, report to Governor Nelson Rockefeller, 1960, p. 7.
3. James E. Carter, Presidential Directive 41, 1978.
4. "Civil Defense Relocation Plan Said to Be Dropped," *The New York Times,* 4 March 1985.
5. Pete Dyke, *The Front Line,* March 1985, Santa Fe, NM.
6. *XI Defense Monitor* (Center for Defense Information, Washington, DC, 1982), No. 5, p. 3.
7. C. Henderson and W. Strape, "Crisis Relocation of the Population at Risk in the New York Metropolitan Area," SRI Project No. 5591, 1978.
8. R. J. Sullivan et al., *Candidate U.S. Civil Defense Programs* (Systems Planning Corporation Arlington, VA, 1978).
9. R. J. Sullivan, "Why We Need CRP," Journal of Civil Defense, August 1982, p. 12.
10. Plattsburgh Crisis Relocation Plans, Office of Disaster Preparedness, Plattsburgh, NY (undated prototype).
11. K. S. Gant and C. V. Chester, Health Phys. **41**, 455 (1981).
12. Editorial, *The New York Times,* 13 November 1978; further criticisms were offered or reported in *ibid.*, 28 December 1978, p. 1 and *The Washington Post,* 5 January 1979, p. A19.
13. *Licensees' Testimony,* Consolidated Edison Company of New York, Inc. (Indian Point, Unit No. 2 (The Nuclear Regulatory Commission, Washington, DC, 1984)).
14. A general summary of public protest against nuclear power can be found in Daniel Ford, *Cult of the Atom* (Simon and Schuster, New York, 1982), p. 4.
15. Dave McLoughlin, in hearings of the Subcommittee on HUD–Independent Agencies, 15 March 1983, Washington, DC, p. 74.
16. *Civil Defense and the Effects of Nuclear War* (Congressional Research Service, Library of Congress, Washington, DC, 1982).
17. Mattie E. Treadwell, *Hurricane Carla, September 3–4 1961* (Office of Civil Defense, Denton, TX, 1961); see also "Civil Defense," hearings of the Senate Committee on Banking, Housing and Urban Affairs, 95th Congress, 2nd Session, January 1979.
18. Cynthia B. Flynn, *Three Mile Island: Findings to Date* (The Nuclear Regulatory Commission, Washington, DC, 1980), Report No. NUREG/CR-1215.
19. T. K. Jones is quoted extensively by R. Scheer, in *With Enough Shovels: Reagan, Bush and Nuclear War* (Random House, New York, 1983).
20. John Kourofsky, in testimony to State Assemblyman Mark Alan Siegel, Plattsburgh, NY, 12 July 1982.
21. *Prototype Basic Crisis Relocation Plan* (Office of Disaster Preparedness, Oneida County, Utica/Rome, NY, undated).
22. John Steinbrunner, Sci. Am. **250**, 1 (1984), p. 37.
23. Herbert L. Abrams, Bull. At. Sci. **40**, 23 (1984).
24. Fred C. Iklé, *The Social Impact of Nuclear War* (University of Oklahoma Press, Norman, OK, 1958).
25. R. K. Laurino, F. Trinkl, R. Berry, R. Schnider, and W. Macdougell, *Impacts of Crisis Relocation on U.S. Economic and Industrial Activity* (Center for Planning and Research, Palo Alto, CA, 1978), Report No. DCPA 01-76-c-0331.

26. Arthur Katz, *Life After Nuclear War* (Ballinger, Cambridge, MA, 1982). See, in particular, Chap. 9.
27. The Hudson Institute has produced many reports on post-attack scenarios. See, for example, W. M. Brown and D. Yokelson, *Final Report: Post-Attack Recovery Strategies* (Hudson Institute, Croton-on-Hudson, 1980), Report No. HI-3100-RR.
28. "Yates Papers," Federation of American Scientists press release, 4 December 1961, box 75.
29. *Nuclear War is National Suicide* (Federation of American Scientists, Washington, D.C., 1981), public interest report, Vol. 34, p. 6.
30. Daniel Graham, press interview quoted in Detroit News, 7 May 1976, p. 513.
31. T. K. Jones, testimony in hearings before the House Committee on Armed Services, 94th Congress, 2nd Session, Civil Defense Review No. 94-42, 1976, pp. 248–252.
32. J. F. Kennedy, message to Congress, May 1961.
33. Ronald Reagan, National Security Directive (Civil Defense), 1982, p. 26.
34. Leon Gouré, *War Survival in Soviet Strategy: USSR Civil Defense* (University of Miami, Miami, FL, 1976).
35. T. K. Jones, in *Defense Industrial Base: Industrial Preparedness and Nuclear War Survival*, hearings before the Joint Committee on Defense Production, 94th Congress, 2nd Session (U.S. GPO, Washington, DC, 1976).
36. W. K. H. Panovsky, in *Civil Defense and Limited Nuclear War*, Congressional testimony before the Joint Committee on Defense Production, 94th Congress, 2nd Session (U.S. GPO, Washington, DC, 1976).
37. H. F. Scott, Air Force Magazine, **58**, 10 (Oct. 1975), and **59**, 8 (Aug. 1976). This publication also has an editorial by J. L. Frisbee on the same topic in the 1976 issue.
38. Leon Gouré, *Shelters in Soviet War Survival Strategy* (University of Miami, Miami, FL, 1978).
39. *U.S. Urban Population Vulnerability* (U.S. Arms Control and Disarmament Agency, Washington, DC, 1979).
40. B. Tuchman, *A Distant Mirror* (Ballantine, New York, 1978).
41. P. R. Ehrlich, Nat. Hist. **93**, 3 (1984).
42. J. Greene et al., *Recovery from Nuclear Attack* (International Center for Emergency Preparedness, Washington, DC, 1979).
43. R. P. Turco, O. B. Toon, T. P. Ackerman, J. B. Pollack, and C. Sagan, Science **222**, 1283 (1983); Sci. Am. **251**, 33 (1984). See also Chap. 8.
44. *The Effects on the Atmosphere of a Major Nuclear Exchange* (National Research Council, Washington, DC, 1984).
45. Edward Teller, "The Aftereffects of Nuclear War," Federal Emergency Management Agency document, 1984 (unpublished).
46. Carl Sagan, address at convocation of United Campuses to Prevent Nuclear War, Cornell University, April 1984.

Chapter 8
Civil defense in other countries

Evans M. Harrell

Introduction

Nuclear war threatens the whole world. In many ways the problems of survival in one land are similar to those confronted elsewhere. For instance, there are the same distinct stages: the long-range preparation period before an attack, the time of tense crisis and warning, the attack itself, the sheltering period after an attack, and recovery. Of course, there are also important differences, and what is politically desirable or possible in one country will not be the same as in another. Still, it would be instructive to compare the United States with the Soviet Union, other NATO and Warsaw Pact countries, minor nuclear powers like China and India, advanced presumptive noncombatants like Switzerland, and poorer non-nuclear countries.

Here I describe the civil defense preparations of the Soviet Union and Switzerland. The Soviet Union, our most likely nuclear adversary, is of interest, for, among other reasons, its civil defense program may have strategic implications. Switzerland, like the Soviet Union, is famous for having a strong civil defense program. The Swiss provide shelter space for virtually the whole population, and are often cited as a model for other countries. Some other advanced neutral countries, including Austria, Finland, Sweden, and Yugoslavia, have programs resembling Switzerland's in many respects. NATO countries vary considerably, with the Scandinavian countries having historically the largest programs, sometimes accounting for over 10% of their military budgets, and France, Greece, and Italy having smaller ones. France, which has not historically paid much attention to civil defense, has recently debated emulating the Swiss, although the chances that France will enact a similar program appear small at present. It is likely that Warsaw Pact countries have programs patterned on that of the U.S.S.R., although information about them is scarce. According to a Chinese civil defense manual,[1] China has a relatively strong civil defense effort with tunnels under its cities. Poorer non-nuclear countries apparently have no civil defense to speak of, although according to some reports they may suffer greatly even if distant from the battleground (see Chap. 9 and Ref. 2).

Some recent representative per capita figures for civil defense expenditures are given in Table I.[3] Bear in mind that comparisons are often difficult because expenditures are allocated differently in different countries and information is not always readily available. The Soviet figure, in particular, is an estimate of what it would cost the United States to duplicate the Soviet program, since actual costs are unavailable.

Civil defense in the Soviet Union

This section is based on the articles and books listed in the bibliography and on conversations with Soviet emigrés. It has also received a critical reading by Federal Emergency Management Agency (FEMA) personnel. Only a small amount of first-hand Soviet material and no classified articles were available. So long as one stays with the factual accounts and away from inference, the picture that emerges from most informed books and articles does not vary excessively. The leading analyst of the subject is Gouré,[4] whose many writings over more than two decades make much use of Soviet sources. He is fairly strongly anti-Soviet and is impressed by the Soviet civil defense programs. His factual account is largely in agreement with those of the CIA,[5] of D. R. Jones,[6] and even Kaplan,[7] who is an outspoken skeptic, but who bases his criticism of Gouré largely on facts cited by Gouré himself. There are a small number of sources, notably testimony before Congress by T. K. Jones and Wigner,[8] which paint an even more frightening picture than these accounts. There are of course many uncertainties owing to the Soviets' penchant for secrecy and their tendency to have grand plans on paper not always matched by practice. In addition, an integral part of the Soviet plans is their emphasis on dual use, i.e., planning everything to serve not only the goals of civil defense but also some other economic or political purpose. If all multiple-use projects are counted as civil defense projects, it is easy to exaggerate the size of the Soviet efforts, and if only the projects that are predominantly for civil defense are counted, then the size will be rather underestimated.

Civil defense in the Soviet Union is overseen by a high-ranking deputy minister of defense, and by all accounts its intended scope is impressive, far beyond what could at present be envisaged in the United States. Less certain are the effectiveness and true purpose of these efforts. Analysts disagree, for example, about how much public apathy interferes with the efforts to keep the public in a state of readiness (to cope with war, or simply readiness to follow orders?) and about how effective the efforts would be in any case. Nonetheless, they all agree that the Soviet programs have some problems in being effective.

For obvious geographical and historical reasons civil defense has been important to the Soviets much longer than to Americans. Already by the 1920s they were actively concerned with civil defense against chemical weapons, and their present-day system is a direct descendent of the one instituted as early as 1932. They have had a sustained, large civil defense effort since that time. Their public instruction program has undergone frequent reorganization over the years in response to changes in the nature of the perceived threats, and possibly also due to dissatisfaction with its operation and a feeling that change would keep it vigorous. Although 137 000 000 citizens were trained in civil defense during World War II, cities such as Leningrad still suffered greatly. The lesson drawn by the Soviet authorities appears to have been that, if anything, more civil defense was needed.

The first concern with civil defense in the context of nuclear war did not appear in Soviet writings until after the Soviets' acquisition of nuclear arms and the death of Stalin in 1953. Evacuation was first mentioned in manuals in 1958 and has been a consistent theme since then. The other elements of their program are large peacetime educational, training, and organizational efforts, elaborate warning systems, construction of fallout and blast shelters for the public, hardening and dispersal of industry, and stockpiling. United States estimates of the U. S. equivalent budget for key elements of Soviet civil defense, mainly personnel and construction costs, have ranged from about a billion dollars per year in the early sixties to about $3 billion in 1982, which is many times what the United States spends. U. S. equivalent estimates refer to the figures that it would take us to exactly duplicate their programs, regardless of the very different allocation of costs in the two economies. Soviet civil defense is labor-intensive, so the fact that our labor costs are so much higher than theirs relative to materials and technology is widely admitted to inflate this figure. On the other hand, some Soviet programs are not included in this figure, so precise quantitative comparisons are problematic.

In 1961 authority for civil defense was transferred from the Ministry of the Interior to the Ministry of Defense, and in 1973 it underwent its most recent reorganization and "reinvigorization," after A.T. Altunin succeeded Chaikov as the Soviet director of civil defense. It is reported that "civil defense has assumed an importance hardly inferior to that accorded the five recognized branches of the armed forces." [6] General Altunin is a Hero of the Soviet Union and a deputy minister of defense. There are at least 60 general officers employed full-time on civil defense at national and subnational levels. "The civil defense hierarchy is theoretically headed by civilian officials and largely funded by civilian agencies, but it is tied together by a military infrastructure controlled by the MOD [= ministry of defense]." [6] There are civil defense organizations at every level of Soviet society, from republic through region, territory, city, and even individual factory. The bulk of the activity is supposed to go on at the grass-roots level, with part volunteer, part professional squads organized at every major industrial and educational concern and every residential district or large living unit. The squads are organized into larger detachments usually of over 100 members.

Civil defense in the Soviet Union is concerned with protecting citizens against not only the effects of nuclear warfare but also those of chemical and biological warfare and, more recently, non-war-related catastrophes such as floods, forest fires, and earthquakes. The parallel movement in the United States is partly intended to persuade the public to accept the cost of civil defense in peacetime, and there may be similar reasoning on the Soviet side. It is also efficient allocation of resources and one of many examples of the philosophy of dual-use. In addition to protection of the public, civil defense in the Soviet Union is very much concerned with the preservation of the military, economic, and political structure of the country, maintaining the ability of the industrial base to continue operating during war, and recovery afterwards. Other generalities cited in Soviet writings as goals of their civil defense are contributing to the "defense capability as a whole," [9] which some[8] interpret as meaning preparation for confrontation with the United States, which would presumably back down if it were obvious that the Soviet people could much better survive a nuclear war, and overcoming "skepticism" about the possibility of survival of a nuclear war. There is ample evidence that such skepticism is widespread. A goal not stated in the Soviet press but believed by many people writing about the issue is the maintenance of a "garrison-state mentality" to keep the Soviet public docile and orderly. Emigrés report in conversation that this is widely believed in the Soviet Union to be a principal goal toward which the training exercises, etc. are quite effective. A related but more charitably expressed desire is said to be that of the Communist Party to reassure the citizens that it is being responsible and doing all it can in the face of the manifest danger.

Testimony before Congress has indicated that whereas U.S. casualties would probably number at least 60% in the event of a full-scale nuclear war, Soviet fatalities might be as low as 5% or even 2%.[8] Few dispute the U.S. estimate in the absence of a beefed-up civil defense program, but considerable doubt has been voiced by independent analysts about the Soviet figure, especially because these estimates depend on the assumption of several days evacuation time. Aspin[10] claims that even with very successful sheltering and evacuation Soviet casualties would be at least in the range of 10% to 20%, and he and others have cited many reasons for doubting the successful implementation of the Soviet plans. The Arms Control and Disarmament Agency estimates that even if the Soviet Union is totally successful in implementing its plans, if the United States were to follow the objective of killing people in retaliation with its reserve weapons, then it can inflict death upon 30% or 35%.[11] (Although this targeting objective is consistent with the doctrine of mutual assured destruction and is commonly raised in discussions of civil defense, targeting of population *per se* has been disavowed by senior U.S. officials, and is felt by many to be militarily absurd.)

The Soviet plans for protecting the public in this possibly formidable way are not so radically different from our own in theory, but are much more intensive and coordinated with the overall planning of the economy. There are five parts to these plans: dispersal, education and training, warning systems, sheltering, and evacuation.

Dispersal refers to the attempt, from the initial planning stage, to ensure that industrial sites and new settlements are either located in less developed areas of the Soviet Union, or at least scattered somewhat widely within a given area. Even the buildings of a single industrial operation will occasionally be spaced widely apart. The civil defense reason for this is simply to make targeting more difficult, and probably has more to do with protecting industry than the public. Essential production facilities are duplicated in dispersed locations, and there are plans for rapid conversion from civilian to military production if necessary. Dispersal serves political and economic purposes at the same time. It is offset to a large degree by the Soviet bureaucracy's fondness for a few large manufacturing centers, like the immense Kama River Truck Plant, rather than many small ones that are more difficult to manage centrally, and by the Soviet people's fondness for living in the big cities, which offer much more in the way of goods, conveniences, and culture. Soviet planned dispersal no doubt has an effect, but probably no more than the unplanned dispersal in our own country due to our decentralized economy and automobile-oriented lives.

The most striking difference between Soviet and American civil defense is in the degree of readiness aspired to and attained through education, training, and overall organization. From 6% to 10% of the population is supposed to be actively engaged in special civil defense squads of 30 or 40 members, organized at a given factory, residence, etc., and subdivided into smaller teams specializing in fire fighting, shelter management, first aid, and so forth. The squads are partially voluntary and partially compulsory. In addition, full-time civil defense professionals number over 100 000. There is a paramilitary organization known by its Russian acronym DOSAAF, with more than a million members and several million adjuncts (estimates of the total number associated with DOSAAF range from 7 to 80 million.[4(e)]) DOSAAF members have received special civil defense training and have been involved in training the general public in civil defense and skills felt to be related, such as radio, marksmanship, and skiing. In addition, the scientific and cultural society Znanie gives frequent lectures on matters related to civil defense among many other topics, and television programs on the subject are often aired.

There have been many compulsory training programs for the entire adult population and school children since the 1950's. These programs are reorganized at intervals of several years or new ones are instituted. The net effect is that every few years every adult must take a civil defense course of between 12 and 22 hours (from 1954 to 1972 the periodic courses required of the adult population totalled 104 hours of instruction, and in 1973 this was increased to 20 hours per year) and must take an examination every year. All workers must take courses and conduct field exercises at their workplaces, and nonworking people are supposed to attend lectures and study privately. There are civil defense courses at the university in addition to military and political training, and schoolchildren study civil defense in second, fifth, and ninth grades (by the end of secondary school a youth has had at least 62 classroom hours and over 50 hours of evacuation and field exercises). Courses cover the nature of modern weapons, first aid, fire fighting, decontamination, shelter behavior, and veterinary aid. Mention has been made in the local Soviet press of apathy towards the educational efforts and even outright fraud, such as plant managers having civil defense crews clear debris or carry out other non-civil-defense tasks while claiming to be conducting field exercises. Although the Soviet authorities are dismayed at widespread problems like this, it still seems that virtually all Soviet citizens are exposed to civil defense training and have at least some minimal theoretical knowledge. Perhaps most importantly, they are familiar with the chain of command that they will have to obey. Although it has been suggested that we should match the Soviet civil defense efforts, including education and training,[8] many Americans would regard this as excessive militarization of our society unless they became more widely convinced that nuclear war was really at hand.

The warning systems instituted are not unlike our own, with air-raid sirens, radio announcements, and so on, the main difference being that the Soviets have a wider variety of systems. With their more widespread training the Soviet citizens can presumably distinguish among more possibilities. Hardening communications systems against disruption by electromagnetic pulse (EMP) has been underway for at least a decade.[12] According to the signals or announcements, the citizens are supposed to evacuate, seek permanent shelters, seek nearby or makeshift shelter, or carry out damage-limiting operations. Different detailed duties are spelled out in advance for those involved at various levels in the civil defense hierarchy, depending on which of nine types of danger are indicated.

The Soviets have built both fallout and blast shelters. The typical shelter is a basement fallout shelter in one of the large apartment dwellings that house most of the urban population. People are supposed to bring enough food and water for several days to these shelters, although some water may be stored there beforehand. Some detached shelters have been constructed, and are supposed to be stocked with food, water, medicine, and other supplies. Shelters are frequently to be used in peacetime as garages, stores, temporary storage areas, study rooms, and even movie houses.[4(e)] Plans have been worked out for putting together additional expedient shelters, if there is time. There are conflicting reports about the availability of blast shelters. There have been ambitious plans to construct them since the 1950s, but in the mid-1960s it was determined that they were too expensive, and construction of them was continued only in the vicinity of factories and important public buildings where essential workers are employed; protection of the leadership has a high priority. There was then some renewal of blast-shelter construction in the 1970s, so that by now there are probably blast shelters for the chain of command and many, if not most, essential workers. In addition, when subways have been built, they have been laid very deep underground so that they could provide effective blast shelters if necessary. The Moscow subway system in particular could in principle accommodate a large number of people (estimates range as high as 10%–20% of the population in Moscow), if properly equipped and stocked. About 75 sites in Moscow have been identified by Western analysts as possible additional blast shelters. One of these turned out on visual inspection to be a public toilet, however.[4(e)] A 1978 CIA report[5] reiterates the claim that 10%–20% of the total urban population can be accommodated by blast shelters, but its assumptions are criticized by Aspin,[10(b)] who points out that they assume only a half to one square meter per person and that sanitation and air filtration are likely to be inadequate, among other problems. Fairly detailed descriptions of the plans for Soviet shelters can be found in the references, especially Ref. 4(b).

There will typically be a five-person civil defense team for every shelter. Shelters accommodate from 100 to several thousand people, and larger shelters have larger teams. The teams are supposed to have access to gas masks, protective clothing, and decontamination equipment as well as the usual stockpiled supplies. It is generally doubted that this equipment is available to people other than essential workers and civil defense personnel.

As in the United States it has been felt that evacuation of the cities is an effective way to save lives, if it can be done. It is the responsibility of the local city soviets (advisory councils) to create evacuation plans. The effectiveness and details of the plans depend, of course, on whether they are implemented before, during, or after a strike. Many inhabitants would be moved only within commuting distance. This is necessary to keep essential industry operating, and because of problems with weather and the condition of the roads and limited availability of private transportation. Indeed, the young and healthy would often be required to evacuate on foot, and even dump trucks would have to be requisitioned to move others. A full evacuation would get under way in a few hours and take three days or more for a large city. There have been calls in the Soviet press for full-scale rehearsals in large cities, but none have been carried out. Reportedly, limited exercises have been carried out in cities with populations up to 40 000, with disappointing results.[6]

The Soviets are concerned with safeguarding industry, beginning with the planning stage by following policies of dispersal, routine hardening such as ensuring that water and gas mains are deep underground, and specific hardening of important sites. A Boeing study has reported that 75% of new industrial construction is being built outside large cities. It has been reported that at least one large industrial site, covering seven or eight million square feet, has been constructed underground,[4(e)] although it is obvious that the vast majority of strategically important sites, such as refineries, are not and probably cannot be protected in this extraordinary fashion. In the event of an alert, workers at essential plants will be divided into shifts which will commute to their plant sites from nearby dispersed blast shelters. Materials and fuels are supposed to be stockpiled, but since supply and distribution are problematic in the Soviet Union in normal times, this is uncertain.

As has already been mentioned, there are many reasons for questioning the effectiveness of the Soviet civil-defense plans, although it is of necessity largely a matter of speculation how serious the problems are. The Soviet press has discussed some reasons for dissatisfaction, such as "apathy" and "frivolity" on the part of the public during training exercises.[4(a)–4(e),13] Conversations both with Soviet emigrés and with people in the Soviet Union have confirmed widespread skepticism and indifference to civil defense. Cynical jokes abound, for example, using the Russian word for coffin, GROB, as the acronym for civil defense, Grazhdanskaya Oborona, rather than the official acronym, GO. There are complaints in the press that instruction is too often purely formal and theoretical and even that there has been wholesale fabrication of testing results. Although the majority of the work force is theoretically involved in civil defense, reports by some recent emigrés suggest that as few as 10% participate in civil defense beyond the compulsory educational courses. D. R. Jones[6] estimates the true figure as about one third. The Soviet press has cited particular difficulties with certain occupations, such as taxi drivers, which are difficult to organize, and has referred to plant managers deliberately scheduling the classes and exercises during busy periods, when they would be forced to cancel. Instruction, especially in rural areas, is uneven and often lacking, due to a shortage of qualified instructors.

Several other problems are less explicit in the Soviet press but not hard to piece together. They include failure to complete construction, scheduling and logistical problems, and chronic equipment shortages. In an apartment-dwelling society even such mundane equipment as shovels could be in short supply when urgently needed to construct expedient shelters. These problems affect all phases of civil defense from education and planning to actual sheltering or evacuation. Dual use makes sense economically but may mean, for example, that a storage room first has to be cleared of its contents before it can be occupied as a shelter. This could be annoying as the bombs are raining down. It is also possible that local bureaucrats may use dual use as a cover for subverting civil defense projects for their own ends. Finally, it is no secret that the Soviet Union often lies under a layer of mud or snow, so weather may always present serious and unpredictable additional obstacles during evacuation.

Some critics have noticed what they feel are serious flaws in specific Soviet plans. Gouré[4(b)] reports that there will be only three days' supply of food and two toilets for each 100–150 people in the average shelter. The shelters will be crowded as well as unsanitary. Oak Ridge scientists feel that the proposed Soviet ventilation systems are inadequate.[7] Moreover, they are often to be powered from outside. Assuming that the United States carries out a policy of retaliatory targeting of the population, and that intelligence and communication networks are undisrupted, evacuation to nearby areas might do little to protect people, when the entire U. S. ICBM force can be retargetted within ten hours.[7] Massive movements of people scheduled to take three days would probably be detected with ample time. Lastly, some studies have claimed that damage unacceptable to the U.S.S.R. can be inflicted even with a small missile force by targeting well-chosen industrial or economic sites, such as their water system, which cannot be saved by civil defense.[14]

What are we as Americans to make of the Soviet civil defense efforts? Should we imitate it either for humanitarian reasons or because it affects the strategic balance? There are two important issues to resolve in order to answer these questions: How much impact does peacetime civil defense have on diplomacy and on the control of international crises? And how much marginal improvement in safety is attainable with increased expenditure, especially within the context of our own society, which is not inclined toward regimentation? Unfortunately both issues are likely to remain to some degree a matter of speculation and argument rather than firm facts.

The Carter Presidential Directive 41 on civil defense,[15] which has been accepted by the Reagan administration, makes reference in a general way to the effect of civil defense on deterrence. Some people have argued that if the Soviets are much better able to survive a nuclear war than we are, then they may be able to push us around in the international arena or even start a war with relative impunity.[8] Now, what really matters in this line of reasoning is the perception, on both sides, of the ability to survive. Western estimates of how well the measures described above would work in practice vary a great deal. The logistical difficulties are tremendous, even with the optimistic assumptions of long warning periods and the like that go into the more impressive estimates. While the Soviets do not publish detailed, impartial analyses of probable effectiveness, we know, on the one hand, that they have the strongest of intentions, but, on the other hand, that they admit to many shortcomings in their locally distributed publications. It is hard to see how there can ever be much certainty as to the effectiveness of the Soviets' civil defense in actual combat, and even the most optimistic assumptions have them losing several million lives and a tremendous amount of industrial capacity. This very uncertainty provides at least some grounds for hoping that the Soviets will never rely on their possible greater ability to survive unless they perceive themselves as having no choice.

One issue frequently raised about civil defense is whether it has a stabilizing or destabilizing effect on the strategic balance. Although the Soviet Union has repeatedly stated that it would not initiate war, in particular nuclear war, against "capitalist aggressors," if faced with evidence of active mobilization, they might well feel compelled to make the first move. The same is an obvious possibility for our side. The question of whether the mere existence of an active civil defense program might be provocative of attack is raised by many of its opponents in the West, and others feel at least that it may aggravate international tension by contibuting to overall war readiness. Soviet military thinkers have never been willing to openly consider the possibility that their own civil defense might have this effect, and seem to regard the necessity for civil defense as obvious. In response to this strictly Western speculation, they have said,[16] "Soviet civil defense does not incite, does not promote, and does not provide impetus to war. Its nature is decisively influenced by the peace-loving foreign policy of the socialist state. Therefore there is no basis for the 'forecasts' of Western experts that a strengthening of the civil defense of the USSR will lead to greater 'inflexibility' of Soviet foreign policy and even to aggravation of international tension." Their feeling that defensive measures should not be regarded as provocative has been consistent since the days of the ABM debate.[17] Moreover, they have never given much credit to the deterrence theory on which this speculation is premised. Some evidence has even been reported[18] that the Soviets might regard American inattention to civil defense as evidence of ignorance about the true dimensions of nuclear war, and thus as a sign of possible recklessness. Further arguments on both sides of this issue are discussed in Chap. 11.

Given the nature of the Soviet system, not everything they say can be taken at face value, and their real motivations have to be surmised. Many of the true reasons for an active civil-defense program may have to do with internal politics, and those in control need not actually believe in the protection of the citizenry to justify action on this basis. They may also be considering other factors than a simple U.S.–U.S.S.R. confrontation with ICBM's. Kincade[14] has speculated that Soviet planners are aware that civil defense would be of little military significance against a massive American attack, but that it might make a big difference against smaller nuclear powers such as China, Pakistan, or England, or in the event that nuclear war is limited to the theater level. Furthermore, he feels that an entrenched civil defense bureaucracy may sustain itself in the Soviet Union despite its increasing uselessness in the face of relentless arms proliferation and technological advances.

It should likewise not be forgotten that Western analyses may be subject to bias (which is not to say that they are necessarily wrong). The study most impressed by the Soviet efforts and implying that the United States should consequently increase its budget for civil defense and arms, for example, was commissioned by the Boeing Company. On the other hand, some strong opponents of American civil defense are clearly partially motivated by mistrust of our own government and political process: "If U.S. leaders believe that our civil defense can be effective, they may be more inclined to attempt to use nuclear weapons for political ends... . Perhaps most important, if Americans believe that they and their nation can survive a nuclear war, the push for 'useable' nuclear superiority would encounter less resistance. An illusion of survival could endanger the movement for a mutual arms 'freeze.' Thus even a facade of civil defense could present an increased danger of war." [19]

This last argument is reminiscent of an early argument against bomb shelters advanced by Dyson,[20] which runs as follows: The building of shelters and concommitant short-term sense of security will eventually lead the superpowers to build even more destructive weapons. This is dangerous for mankind, because with more weapons and more destructive types of weapons, the point will eventually be reached where noncombatant nations will die as well as the involved populations. In the end, rather than local calamities, there might even be global extinction. As chronicled by Kincade[14] and Weart (Chap. 2), there have been several outbreaks of civil defense fervor in the United States since the Korean war, each of very short duration, because Congress has never felt that a large-scale civil defense program would either work well or be worth the money. Sadly, it appears that the arms race goes on whether or not the public has any sense of security.

Swiss civil defense

Switzerland has attracted much attention for its strong civil defense program, which attempts to offer shelter space for the whole population. The number of inquiries from around the world is so great that the Federal Office of Civil Protection in Berne has resorted to issuing a form letter in several languages about Swiss shelters.[21(c)] The Swiss program offers many interesting contrasts both with our own and with Soviet civil defense. One obvious contrast with the Soviet Union is the availability of information. Switzerland is an open society. Its publications are readily available, often in English translation provided by the Swiss authorities, and budget figures, shelter designs, and historical and legal accounts are easy to obtain. Nothing prevents the casual tourist from verifying the presence and nature of Swiss shelters or picking up pamphlets in supermarkets. In some ways information is even more available in Switzerland than in the United States. For example, the catalog of the Forschungsinstitut für militärische Bautechnik (Research Institute for Military Construction Engineering),[22] which publishes most of the government reports connected with civil defense, lists many reports on the generation and effects of electromagnetic pulse, some apparently with more detail than what is readily available in the United States.

Switzerland has a long tradition of neutrality, which has been a permanent state of affairs since the Treaty of Paris (1815). Yet the Swiss never abandoned the fine military tradition that made them so sought after for centuries as mercenaries, and neutrality for the Swiss means armed neutrality. The seriousness of this sentiment was shown in 1964, when a national referendum renouncing the acquisition of nuclear arms lost. Some 10% of the population serves in the army, and the military budget, at 2.3% of the gross national product, while not proportionately very high in comparison with many countries, is reasonably large considering the wealth of the country and the absence of an arms race, hostile neighbors, or foreign adventures. (Some figures for comparison are Japan 0.9%; Canada, 1.9%; ltaly, 2.6%; West Germany, 3.6%; U.S., 6%; U.S.S.R., 12%.[21(b)])

Swiss civil defense traces its origins to 1934, when the Passive Defense Troops were organized. The modern era began in 1959, when a civil defense article was voted into the constitution by a strong majority of the voters, and in 1962 a Federal civil defense law was passed requiring compulsory civil defense service from every healthy male aged 20 to 60 not drafted into the military, and allowing women over 16 to volunteer. Registration began in 1965. At present there are at least 500 000 men doing compulsory civil defense service as well as 20 000 volunteer women, which could be compared to the size of the army, 650 000. Switzerland's population is about 6 500 000, somewhat less than that of New Jersey. Of these troops, about 300 000 have undergone "full civil defense training" at one of the country's 57 centers (at least 10 more are being built). Each person joining civil defense has a five-day introductory course, followed by two days of exercises per year, while higher grade personnel exercise up to twelve days. People in civil defense service are entitled to pay and insurance benefits.

In 1963 a Federal law put an ambitious shelter-building program into effect and set up a Federal Office of Civil Protection. However, in the 1960's there was a feeling that the Swiss civil defense program did not as yet adequately address the threats of modern warfare, and in 1966 the Department of Justice and Police convened a Committee for Civil Defense to carry out detailed studies over the next few years. These studies culminated in a document, The 1971 Conception of the Swiss Civil Defense,[21(a)] which was approved by the Federal Council and submitted to Parliament. It became the cornerstone of Swiss civil defense as over the next several years the civil-defense laws were comprehensively reviewed and modernized. Even though at that stage international tensions were apparently decreasing, the Swiss held to the long view that so long as the Eastern and Western alliances are heavily armed with nuclear weapons, their "conflicts can always contain a germ of nuclear war." Particularly since a shelter-building program takes decades to complete, they felt that preparations should be made continuously.

One of the main conclusions of that study is in striking contrast with either American or Soviet thinking about protecting the public. The Swiss report regards relocation as unworkable, and therefore recommends concentrating totally on a program to provide the whole population with blast and fallout shelters. This is relatively expensive, and construction accounts for most of the Swiss civil-defense budget. In arguing for shelters rather than evacuation the report states that[21(a)] "Modern methods of mass destruction, especially their employment with the element of surprise, practically forbid in our country the possibility of evacuating the population into 'safe' areas.

The deployment of arms of massive destruction from the air or from a neighboring country can endanger all regions of our country, even the thinly populated areas. It would not be possible to guarantee the transfer of the population and their victualling at the reception centers, during War operations. Furthermore, such evacuation might hinder important actions undertaken within the scope of national defence...evacuations on a large scale are ineffectual and even dangerous for Switzerland."

The report does not dwell on contrasts between Switzerland's program and those of other countries, although it considers a range of threats, the probabilities of which are clearly different for a neutral country than for a NATO or Warsaw Pact country. Many discussions of civil defense focus on the scenario of a fairly large exchange between the superpowers or between the Eastern and Western alliances, and in that eventuality Switzerland would have as its primary immediate threat fallout from attacks on neighboring countries. There would then indeed be no point in relocating to the countryside rather than to fallout shelters at people's usual places of residence. It is interesting that this reasoning is not prominent in the 1971 Conception; fallout from war operations in neighboring countries is mentioned only briefly under "Other Dangers." Taking note of the large stockpiles of weapons of the alliances, the Committee speaks of Switzerland's being involved in a possible war as a participant or at least as a military corridor. It also refers to the possibility of being bombed by error, without explicitly calling to mind that this was Switzerland's unhappy experience in the last World War. The range of threats considered includes not only full-scale nuclear warfare, but also conventional, biological, and chemical warfare; limited nuclear war; nuclear blackmail by a hostile power; and accidents and natural disasters.

The report enunciates several guiding principles for civil defense thinking in Switzerland. Those for the shelter program are the following. (1) All inhabitants of the country should have access to a shelter space. (2) Although one must "forget the old idea of being able to give sufficient warning of attack," a crisis period is regarded as likely before an actual nuclear attack. During this period there would be "preventive and gradual occupation of shelters." (3) Shelters should provide for a stay of up to several weeks. (4) The shelters should be of simple and sturdy design. (5) The population will not be evacuated. (6) Finally, there should be a diversity of construction plans, the better to cope with unforeseeable developments affecting the efficacy of any given design. Principles of a more general nature include the observation that "Absolute protection is impossible." Thus the shelters should provide a certain measure of blast protection (1 to 3 atmospheres of overpressure), but should not aspire to the ability to withstand a direct hit. There should be a great deal of flexibility in civil defense to cope with all possible threats. Some of the general principles are reminiscent of trends in American and Soviet civil defense, i.e., the "dual-use" of materials and installations, and the integration of war-related civil defense with management of other emergencies. Switzerland being Switzerland, the committee did not neglect to consider financial planning for the civil defense program.

The main outcome of the 1971 recommendations was a continuation and increase of the building of shelters, with the goal of providing every inhabitant with a high-grade shelter by the 1990s. In the words of Dyson, "The quality of Swiss shelters is even more impressive than their quantity.... Swiss shelters are massive reinforced-concrete structures built into the foundations of buildings."[23] According to the Federal Civil Protection Office, the typical shelter is in a corner of a below-ground basement, is built of reinforced concrete with a 40-cm-thick ceiling, and has a ventilating system and dust filter. Switzerland has developed several designs of shelters of all sizes, both public and private, as well as designs of protected command posts and the like [floor plans are reproduced in Ref. 21(b)]. The price per space, 1–2.5 m^2 a person, was estimated as from 425 Swiss francs in a large shelter for 100 people to 1150 Swiss francs for a single-family shelter in 1978.[24] A somewhat higher figure, 25 000 Swiss francs (1983) for 5–7 spaces, is quoted by the Federal Civil Protection Office in its form letter in response to foreign inquiries.[21(c)] (In the intervening time the franc fell from about $0.60 to about $0.40, and any additional discrepancy can be accounted for by inflation and by the assumption in the letter to foreigners that materials like armored doors would have to be imported from Switzerland.) Costs could be cut substantially by modifying preexisting underground rooms, but, on the other hand, they would be about doubled in the absence of a basement in which to build them. Apparently, almost all Swiss housing comes with a basement.

The Swiss laws have explicit financial guidelines and cost-sharing arrangements for the construction of shelters. The added cost of a building because of the addition of a shelter shall not exceed 5% of the total cost, and it is shared 50-50 by the Federal government. Public shelters are built in areas where there are not enough private shelters, and come equipped with underground car parking. They are a shared responsibility, with 70% of the cost borne by the locality and 30% by the Federal government. Because of the cost-sharing, construction accounts for about 82.5% of the total Federal expenditures for civil defense (17% goes for equipment and training, and the rest is for research and development). By 1981, 75% of the inhabitants of Switzerland were provided with fully protected shelter spaces, and 100% had at least makeshift shelter. The government provides explicit instructions for stockpiling of emergency supplies and periodically inspects shelters.

In addition to personal shelters, the Swiss have built protected command posts (1000 of an eventual ca. 2000), preparation facilities (570 of ca. 1500), and underground first-aid stations and hospitals. Preparation facilities are where material and equipment are kept for emergency distribution. The underground hospitals are usually built in conjunction with ordinary peacetime hospitals. There are over 78 000 protected hospital beds (of 130 000 planned), 715 first-aid posts, 279 first-aid stations, and 97 emergency hospitals and operating rooms. First-aid stations are larger and more elaborate than posts, having pharmaceutical stores, and operation and morgue rooms. About 70% of the planned civil defense material, presumably meters, communication equipment, etc., are reported to have been delivered. (Figures are as of January 1, 1982.)

There is little or no discussion in the Swiss literature of the ability of civil defense to enhance war-fighting ability, protection of critical military industry, and preservation of authority. This tacitly reflects the expectation that

Switzerland will not be a primary combatant in a likely nuclear war.

Like the American and Soviet civil defense programs, much responsibility is delegated to the local level, in this case the community or municipality. Both the Cantonal and Federal governments issue regulations and participate in planning, but the local government is charged with actually carrying out all the civil-defense plans. There are three civil defense organizations at the local level: the local protective organization, the industrial defense organization, and the shelter organization. All of them report to the local director, who is responsible to the communal council. In addition, in time of emergency the Air Defense Troops can be placed at the disposal of the civil defense organizations. The shelter organization readies the shelters and takes in the homeless, while the local protective organization is charged with general preparation, alarms, rescue, fire fighting, etc. (Detailed breakdown of the division of authority is to be found in Ref. 25.)

Why the Swiss? Why should this neutral, central European country have quite possibly the strongest commitment to civil defense in the world? Swiss citizens, like Russians, have been known to turn the question around and express surprise at the lack of commitment of so much of the rest of the world, particularly of countries devastated in the World Wars. Historical accidents are not impossible, and it could simply be that the debate on the merits has gone differently in Switzerland and the other countries that have developed similar programs than in so many other countries. There are, however, aspects of the Swiss tradition and national character that may account for some of the differences in attitude. The Swiss are perceived by other Europeans as staid, religious, careful, and conservative. They are also felt to mistrust foreigners. They have long been known for their devotion to health and safety. It is perhaps no accident that the country that founded the Red Cross now has prepared for nuclear warfare by building underground first-aid stations and hospitals. Another remarkable fact about Switzerland is its wealth. It is the second richest country in the world (after Kuwait), and the disparities in wealth among Swiss citizens are much smaller than in the United States. Very few members of the voting public would find the Federal expenditure of roughly $33 per year per capita burdensome.

The Swiss civil defense program is not without its critics. The booklets published by the Federal Office of Civil Protection are quite frank about the objections that have been raised, which are much like those raised in the American debate on the subject (with less emphasis on whether civil defense is destabilizing).[25] The Physicians for Social Responsibility (PSR) and the International Physicians for the Prevention of Nuclear War have attacked the Swiss civil defense program, for instance in an article by a Swiss doctor, Lauterburg.[26] PSR has studied the case of a 1 Mt bomb dropped on Berne, the capital of Switzerland, with a metropolitan-area population of 400 000. They claim that with no shelters the immediate casualties would be 200 000 dead and 80 000 wounded, whereas with shelters the figures would still be as high as 130 000 and 40 000, which they regard as not good enough to justify the expense or to deter international "blackmail." Lauterburg also charges, "The shelters in question are not equipped appropriately, and a delay of a week has to be expected before people have taken up residence in them." Since it can hardly take a week to enter a shelter in one's basement, the latter remark is presumably a reference to the announced policy of gradual occupation in a time of mounting international crisis. Other arguments are reminiscent of those made in the United States and elsewhere: that civil defense does not do much for long-term survival and recovery, that resources could be better used in other ways, etc. On the other hand, Dyson, a long-time opponent of building bomb shelters in the United States,[20] in a recent publication[23] draws a distinction between Swiss and American shelters. He regards American opposition to civil defense and Swiss adherence to shelter building as both morally correct for the two different countries; the points of distinction are the United States' status as a probable belligerent and the currently inequitable distribution of what meager shelter capacities we currently have. In contrasting the two countries, he says, "If the United States ever wishes to build shelters in a serious fashion, two preconditions are essential: equal right of access to shelters for everyone must be established as a legal principle, and the doctrine of Assured Destruction must be abandoned. Even after these conditions are fulfilled, it is unlikely that Americans will want to spend large sums of money on civil defense." Yet he goes on to point out that in comparison with modern weapons systems, particularly ABM systems, "shelter systems are, in the jargon of the experts, cost-effective" for saving lives.

Table I. Per capita expenditures for civil defense.

Country Cost in U.S. $	France	U.S.	U.K.	Italy	Denmark	U.S.S.R.	Switzerland
	0.15	0.75	1.15	2.00	6.50	11.30	33.00

Conclusions

It is difficult to assess the implications of Soviet and Swiss civil defense for the United States. Since the problems of logistics, design of shelter, etc. are not too dissimilar for the U.S., the U.S.S.R., Switzerland, etc., foreign countries' plans are bound to be of interest to American civil defense professionals and strategic thinkers. On the other hand, their political and social traditions differ quite a bit from ours. In particular, we would find it difficult to mobilize to the extent of either the U.S.S.R. or Switzerland. Hence their actions may not be a reliable guide to American policy decisions.

Postscript: Civil defense during the Chernobyl disaster

As we go to press, the aftermath of the radioactive disaster at the Chernobyl nuclear power station in the Ukrainian S.S.R. is affording a rare glimpse of Warsaw Pact civil defense in action. At this stage the sequence of events is still not known with certainty, but it appears that a partial core meltdown occurred in Reactor Number 4 at 1:23 a.m. on April 26, 1986. An explosion, probably due to the formation of hydrogen from water in contact with overheated fuel, blew the roof off the reactor, which had no containment vessel, and the graphite core burned for many days until it was controlled by firemen and emergency workers, mainly by dumping sand, clay, and boron on the reactor from aircraft. Radioactive material was lofted into the atmosphere and carried by an unusual wind pattern to Scandinavia, where fallout was detected on a worker at a Swedish nuclear power plant by routine monitoring and traced to the Soviet Union. The Soviet government acknowledged the disaster on the evening of April 28. Two workers were killed in the accident, and 299 people have been hospitalized for radiation sickness, of whom so far 26 are dead and dozens more are in serious condition. 18 000 have received some treatment. The Soviet government formed a commission to prepare a detailed report on the accident, which is now available to outsiders.[33] If the commisson turns to civil defense, it will have the best opportunity to study the effects of fallout since the banning of above-ground bomb tests: According to an article which, incredibly, profiled the Chernobyl reactor shortly before the accident as "safer than driving a car," there have been extensive tests on "the flora and fauna, the air and water" to make certain that the station was "ecologically pure" both before the plant was built and at intervals since then.[27]

The early reports in *Pravda* (6, 23, 26 May, 4 June)[28] consisted primarily of charges of unfair treatment by the Western press and governments, recounting of Western nuclear incidents, and praise for the heroism of Soviet emergency workers, rather than factual reporting as an American reader would understand the term. Yet by Soviet standards, where the tradition is not to mention disasters at all, the amount of reporting has been unprecedented, and it shows the seriousness of General Secretary Gorbachev's campaign for openness ("glasnost"). There has even been live coverage of Chernobyl on Soviet television. On April 29 the United States maneuvered a KH-11 military reconnaissance satellite into a position to look at the reactor, and Western reporters began contacting people in the Ukraine, Byelorussia, and Poland. Thus, after some initial wild speculations, unfortunately given credence by Kenneth Adelman, Director of the Arms Control and Disarmament Agency, the Western news media have had access to their own sources of information. What picture of Soviet civil defense is painted by Soviet and Western news accounts?

First, even though the circumstances were vastly more favorable than would be expected in a nuclear war, there have evidently been several foul-ups. No evacuations were undertaken for over 36 hours after the accident, when the 25 000 residents of the town of Pripyat moved out with the aid of thousands of volunteers and 1100 buses from Kiev, an operation taking about 3 hours. This was supposed to clear a zone of 10 km around the reactor of all people other than emergency workers and workers managing the other reactors while they were being shut down. Reportedly, however, an American reconnaissance satellite saw people playing soccer inside the fence of the burned-out reactor two days afterwards.[29(a)] Eventually the evacuation zone was increased to a radius of 30 km from the reactor, requiring the relocation of about 100 000 people, many of whom will never be able to return—new homes and barns for these people are already under construction. Alexander Lyashko, the premier of the Ukraine, was quoted as saying that "people living from six to 18 miles away weren't told to leave until a week after the disaster."³⁰⁽ᵃ⁾ The three officials in charge of evacuating Pripyat have been disciplined.

Reports of the levels of radiation near the plant vary widely: *Newsweek*,²⁹⁽ᶜ⁾ quoted B. Yeltsin, interviewed on West German television as saying that radioactivity in the area "had declined to 200 rem an hour," but other reports have cited measurements of a few rems, and Soviet officials were quoted as saying that "radiation within the 18-mile evacuation zone around the plant peaked at 10 to 15 millirems per hour." ³⁰⁽ᵇ⁾ These discrepancies do not indicate an absence of monitoring equipment, since radiation has clearly been measured not only at Chernobyl but at many sites in the Ukraine, Byelorussia, and elsewhere. It is possible that the radioactive plume carried material into the atmosphere rapidly enough that ambient levels of radioactivity on the ground were less than might be expected. Emergency workers wore special clothing enabling them to work short periods at the plant to bring it under control. The Soviet report estimates a total release of 50 million curies.³³

The citizenry apparently remained calm and cooperative for the most part, although some hundreds of residents of the large nearby city of Kiev evacuated spontaneously, or at least sent their children away from the region. *Tass* quoted Anatoly Romanenko, the Ukrainian health minister, as saying also that some panic-prone people took "medicines that were alleged to protect them from radiation, and there were cases of poisoning. They are now being treated in hospitals." ³¹ The Soviet press also stated that some workers at Chernobyl deserted their posts in panic, and that some people had refused to take in evacuees. The Western press has emphasized these shortcomings more than the heroism that figures so much more prominently in Soviet accounts.

As to civil defense outside the evacuation zone, we learn that residents of Kiev were given straightforward advice to "wash their hair and hands and their floors every day," ³⁰⁽ᵃ⁾ as well as warnings against eating leafy vegetables and staying out of doors a long time. Food and people in the Kiev region have been checked for radiation. In many places in neighboring Poland, iodide was administered to all children up to age 16, showing that at least some such emergency supplies must have been stockpiled in the Warsaw Pact countries. Polish scientists have told me that they and most of their colleagues found it easy to obtain iodide in Warsaw.

The Soviet Union will find it only prudent to try to learn from its mistakes and improve future civil defense. In this regard it should be recalled that the accident at Chernobyl may have been only the second largest radioactive disaster to have occurred in the Soviet Union, after the mysterious incident at Kyshtym, in the Southern Urals, in the winter of 1957. A large region, estimated as "no less than fifteen hundred square kilometers," was radioactively contaminated, and 200 000 people are said to have been relocated (Ref. 32, pp.73, 168). The Soviet Union has never acknowledged the incident, but there are many confirming reports by travelers in the region of signs on the roadside warning people not to stop in the contaminated region. According to Medvedev's reconstruction, there was a (conventional) explosion at a site where radioactive wastes from the Soviet nuclear weapons program were being stored. As to civil defense Medvedev states (Ref. 32, p. 21), "The first seriously organized evacuation was begun after several days, and then only in the settlements closest to the site of the explosion. Subsequently, symptoms of radiation sickness began to appear in more distant areas.... The evacuation affected several thousand persons, possibly tens of thousands, but the number who died of radiation sickness remained unknown."

Simple civil defense measures such as eliminating contaminated produce were undertaken. Obviously, there was not enough disruption to attract the eyes of the Western press at the time, but hospitals in the area were filled with victims, and there was supposedly some panic in Kamensk-Uralskiy, Sverdlovsk, and Chelyabinsk (Ref. 32, p. 133). It appears, in short, that there were many similarities between the two known times that Soviet civil defense was called upon during major civilian radioactive disasters. Both sites were rural, but would be counted as "high-risk" areas, since even the reactor at Chernobyl was producing plutonium for the Soviet military. They might therefore be expected to have benefited from emergency planning beforehand. Although evacuation eventually saved many people, there were serious delays and a certain amount of confusion and panic.

Other sources of information on Chernobyl

(1) Associated Press News Summary of the Accident at Chernobyl, via Compuserve, as of May 9, 1986.

(2) "Soviets Reveal Flight of Nuclear Workers," *Atlanta Journal and Constitution* (from Associated Press), 18 May 1986, p. 4A.

(3) "Some Chernobyl Refugees Might Not Return," *Atlanta Constitution* (from Associated Press), 24 May 1986, p. 4A.

(4) National Public Radio "All Things Considered," broadcasts of 6,8,13 May, and 3,4 June, 1986.

(5) "Reactor Explodes Amid Soviet Silence," Science **232**, 814 (1986).

(6) "Escape Snafus," *Wall Street Journal*, 9 May 1986, p. 1.

(7) "Soviets Report 6 Deaths from Accident at Chernobyl and Discipline 3 Officials," *Wall Street Journal*, 13 May 1986, p. 4.

(8) "Experts Suspect Explosion but Differ About Meltdown," *Washington Post*, 30 April 1986, p. A17.

(9) B. G. Levi, "Cause and impact of Chernobyl accident still hazy," *Physics Today* **39**, No. 7, 17 (1986).

References

1. *Chinese Civil Defense*, edited by C. V. Chester and C. H. Kearny (U.S. GPO, Washington, 1974), Oak Ridge National Laboratory Technical Report No. ORNL/tr-4171. This is a translation of a Chinese civil defense manual. It does not contain much analysis of the actual status of Chinese civil defense. Available from the National Technical Information Service.

2. *The Aftermath: The Human and Ecological Consequences of Nuclear War*, edited by J. Peterson (Pantheon, New York, 1983) [Originally published as Ambio II, Nos. 2–3 (1982)]. The major reference on global effects of nuclear war, with implications particularly for third-world civil defense.

3. "Program History," Federal Emergency Management Agency internal document dated August 1983.

4. L. Gouré, (a) in *The Utilization of Fallout Shelters*, edited by George W. Baker, John H. Rohrer, and Mark J. Nearman (National Academy of Sciences, National Research Council, Washington, DC, 1960); (b) *Civil Defense in the Soviet Union* (University of California Press, Berkeley, 1962); (c) *Soviet Emergency Planning* (Rand, Santa Monica, 1969), Technical Report No. P-4042; (d) *Soviet Civil Defense Revisited, 1966–69* (Rand, Santa Monica, 1969), Technical Report No. RM-6113; (e) *War Survival in Soviet Strategy, U.S.S.R. Civil Defense* (Center for Advanced International Studies, Miami, 1976); (f) Bull. At. Sci. **34**, 48 (1978). Gouré is the person most familiar with primary Soviet sources.

5. *Soviet Civil Defense*, Federal Emergenc;y Management Agency Report No. FEMA-52, 1983 (U.S. GPO, Washington, DC 1983). This report is based on U.S. Intelligence reports and is said to be substantially the same as the *CIA* study NI-78-10003. It is reasonably balanced in its presentation.

6. D. R. Jones, Sov. Armed Forces Rev. Ann. **2**, 289 (1978). Jones, no relation to T. K. Jones, is somewhat skeptical of the effectiveness of Soviet civil defense.

7. F. M. Kaplan, Bull. At. Sci. **34** No. 3, 14 (1978); **34** No. 4, 41 (1978), part 2. Kaplan feels that fears of Soviet civil defense are greatly exaggerated.

8. *United States and Soviet Civil Defense Programs*, U.S. Senate Foreign Relations Committee Hearings, 16 and 31 March 1982 (U.S. GPO, Washington, DC, 1982). Statements by Admiral N. Gayler, T. K. Jones, S. M. Keeny, Jr., R. N. Perle, E. P. Wigner, Brigadier General W. J. Doyle, L. Gouré, J. Leaning, and E. Chivian (International Physicians for the Prevention of Nuclear War).

9. N. V. Ogarkov, "Toward a War Footing," *The Sun* (Baltimore), 9 July 1982, p. A15. Excerpt from a Russian pamphlet entitled "Always in Readiness to Defend the Homeland."

10. L. Aspin, (a) "Soviet Civil Defense: Myth and Reality," *Arms Control Today*, September 1976; (b) Bull. At. Sci. **37** No. 2, 44 (1979), review of the report entitled "Soviet Civil Defense" by the Director of Central Intelligence, Report No. NI-78-10003, July 1978.

11. *The Effects of Nuclear War* (Allanheld, Osmun and Co., Totowa, NJ, 1980), issued by the Office of Technical Assessment, 1979.

12. (NPR): "All Things Considered," National Public Radio broadcast 29 April 1985.

13. D. I. Mikhaylik, Voennie Znaniya **12**, 10 (1983).

14. W. Kincade, Int. Security **2**, 99 (1978). An interesting historical perspective.

15. James E. Carter, Presidential Directive 41, September 1978.

16. Harriet Fast Scott, Air Force Magazine **58**(No. 10), 29 (1975).

17. J. R. Schlesinger, *Arms Interaction and Arms Control* (Rand, Santa Monica, 1968), Technical Report No. P-3881.

18. R. Ehrlich, *Waging Nuclear Peace: The Technology and Politics of Nuclear Weapons* (State University of New York Press, Albany, 1985).

19. J. Lamperti, Bull. At. Sci. **39** No. 6, 7S-10S (1983).

20. F. J. Dyson, Bull. At. Sci. **18** No. 3, 14 (1962).

21. (a) *The 1971 Conception of the Swiss Civil Defence* [Bundesamt für Zivilschutz (Federal Office of Civil Defence), Bern, 1971; (b) *Civil Defence (Civil Protection) Figures Facts Data, 79/80* (Federal Office of Civil Defence, Bern, 1979) a one-sheet enclosure contains 1982 updates of some of these Swiss and world statistics; (c) Untitled form letter, Federal Office of Civil Defence, Bern, dated 1983/84.

22. *Verzeichnis der Veröffentlichungen* (Forschungsinstitut für militärische Bautechnik, Zürich, 1980). A catalog of Swiss civil defense materials, especially technical reports, most in German, with some in English or French.
23. F. J. Dyson, *Weapons and Hope* (Harper and Row, New York, 1984).
24. P. Piroué, *Civil Defence in Switzerland* (Bundesamt für Zivilschutz, Bern, 1982).
25. R. Aeberhard, *The Swiss Civil Defence 81/82* [Bundesamt für Zivilschutz (Federal Office of Civil Defence), Bern, 1981], a free booklet describing the Swiss program.
26. W. Lauterburg, "The Civil Defence Program in Switzerland," International Physicians for the Prevention of Nuclear War Report, Vol. 2, No. 2, p. 12 (1984).
27. "Born of the Atom," *Soviet Life*, February 1986, p. 13.
28. Yu. Zhukov, "Nevol'noe Samorazoblachenie" (Involuntary Self-revelation), *Pravda*, 6 May 1983, p. 4; V. Yavorivskiy; "Zona Pravdy i Sovesti" (Zone of Truth and Conscience), *ibid.*, 23 May 1986, p. 6; E. Parnov, "Cherno-beloe i tsvetnoe" (Black-white and Colored), *ibid.*, 26 May 1986, p. 6.
29. (a) "The Chernobyl Syndrome, *Newsweek*, 12 May 1986, p. 22; "The Anguish of Mikolajki," *ibid.*, 12 May 1986, p. 35; (c) "The 20th-Century Plague," *ibid.*, 12 May 1986, p. 36.
30. (a) "Soviets Inquire on Buying Food From the West," *Wall Street Journal*, 9 May 1986, p. 1; (b) "Soviet Workers Trying to Seal Reactor's Core," *ibid.*, 12 May 1986, p. 4.
31. "Children are Evacuated from Kiev as Wind Shifts," *Atlanta Constitution* (from wire reports), 8 May 1986, p. 4A.
32. Z. A. Medvedev, *Nuclear Disaster in the Urals*, translated by George Saunders (W. W. Norton, New York, 1979).
33. "Chernobyl: Errors and Design Flaws," Science **233**, 1029 (1986).

Chapter 9
Civil defense implications of nuclear winter

Barbara G. Levi

Introduction

The expected effects of nuclear weapons vary with both distance and time—from the prompt gamma rays that could irradiate an area of a few square miles within the first second, to the stratospheric fallout that might descend on the globe over the subsequent months to years. Different civil-defense measures are appropriate to these various weapons' effects. Among the farthest-reaching and longest-term effects of a large-scale nuclear war is the possibility that it might induce a "nuclear winter." The term "nuclear winter" describes a theory according to which Earth's normal climate might be greatly perturbed if the smoke from fires begun by nuclear explosions and the dust raised by ground-burst nuclear warheads block out the sunlight. The soot strongly absorbs incoming sunlight but transmits most of the outgoing infrared radiation, creating an inverse greenhouse effect. The northern hemisphere, where such a war is most likely to occur, might then suffer long periods of darkness and cold, with temperatures returning to normal values only after perhaps a year or more. At the very least, the perturbations caused by the presence of large amounts of dust and soot might introduce erratic fluctuations in the normal weather patterns.

This prediction is so recent, and currently based on so many uncertain factors, that it is still difficult to gauge the possible implications for civil defense. If true, it would greatly complicate the already challenging tasks of survival and recovery. This chapter summarizes the predictions of the nuclear winter models, stressing the key factors in producing the climate effect, and underscoring the major uncertainties in the predictions. It also explores what might be the possible implications for civil defense.

Predictions of the nuclear winter models

The calculations that coined the term nuclear winter were done by a team of scientists with the acronym TTAPS (Turco, Toon, Ackerman, Pollack, and Sagan).[1] They examined quantitatively an idea that had been presented more qualitatively a year earlier by two atmospheric chemists.[2] Although scientists have long conjectured that the dust raised by many nuclear ground bursts might cause the average temperature to fall on the order of 1 °C during

Table I. Nuclear winter baseline scenario (from Ref. 1).

Total Weapon yield (Mt)	Height of burst	Target type	Emissions to atmosphere (Tg[a])	
			Soot	Dust
1000	Air burst	Urban, industrial	149	0
1150	Air burst	Military targets		0
2850	Ground burst	Military targets	80	960
5000			229	960

[a] One terragram (Tg) = 10^{12} g.

the subsequent year,[3] Crutzen and Birks suggested for the first time that the smoke from fires started by nuclear bombs might cause a much more dramatic temperature drop.

The TTAPS team examined many different nuclear-war scenarios but took as their baseline case a nuclear war involving 5000 Mt—roughly half of the yield in the strategic arsenals of the U.S. and U.S.S.R. today. For this baseline scenario, they assumed that warheads with a total yield of nearly 3000 Mt might be ground burst in order to destroy hardened military targets such as ICBM silos. Another 1000 Mt or so was taken to be airburst over other, nonurban military installations and the remaining 1000 Mt was assumed to be directed against urban, industrial targets. The total amounts of dust and smoke resulting from this scenario are shown in Table I. A study by the National Academy of Science[4] formulated a similar scenario for its baseline. Note the strong influence of city fires: Although the warheads directed at cities comprise only 20% of the total yield, they produce about 65% of the total smoke.

To study the climate impact of the dust and smoke, TTAPS simulated the radiative and convective processes occurring in the atmosphere with a one-dimensional model. The one dimension in their model was the height above Earth's surface. They were forced to assume that the smoke was spread uniformly over the northern hemisphere and were unable to treat the moderating influence of the ocean, with its large heat capacity. The results for their baseline is shown as the solid dark curve in Fig. 1 (based on Ref. 1). That figure shows the predicted average temperature over northern hemispheric land masses as a function of time after the hypothetical nuclear conflict. For the baseline, the temperature falls by 35 °C within 20 days and still remains below normal after nearly a year. The TTAPS group estimates that the buffering effect of the oceans might make the temperature drops smaller than those shown in Fig. 1. For example, the temperature might fall by at most 24 °C at the centers of the continents and by 10 °C along the coasts, rather than by 35 °C, as suggested by Fig. 1.

By examining the results for other scenarios in Fig. 1, we can get an understanding of the different effects of the dust and the soot. The scenario labeled "low-yield airbursts" reflects the impact of smoke with no dust, while that labeled "dust only" represents the impact of dust with no smoke. Comparison of the two curves indicates that, in this model, the dust produces a

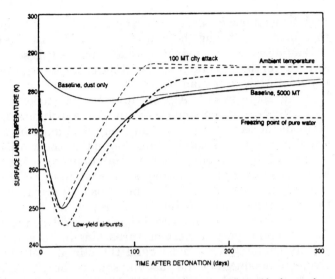

Figure 1. Predicted changes in average temperatures over continents in the northern hemisphere as a function of time after various nuclear attack scenarios. In the baseline (solid light curve), one-fifth of the 5000 Mt is targeted on cities. The air-burst case (broken light curve) includes more cities, which the authors feel will produce more soot and result in greater solar attenuation. Solid heavy curve represents the baseline but with no fires set: The temperature drop is less but the dust prolongs the cooling. Broken light curve illustrates a 100 Mt attack exclusively on city centers, where the density of combustion materials is assumed higher than that in the baseline scenario. (Figure from Ref. 5 and based on Ref. 1.)

small but prolonged cooling while the smoke causes a much sharper but shorter-lived temperature drop.

A fourth curve in Fig. 1, labeled the "100 Mt City Attack," indicates the key role that is played by city fires. TTAPS estimates that, in the innermost 5% of the area of a given city, the density of material that might burn in a fire is perhaps five times higher than the dry biomass content of forests. Furthermore, if a nuclear bomb were to strike such a city center, the fire is expected to be sufficiently intense to burn virtually all the fuel, as it did in the center of Hiroshima. By contrast, only 25% or less of the biomass is consumed in a typical forest fire. Hence, as pointed out above, fires in urban areas can dominate the contribution of soot to the atmosphere.

The 100 Mt scenario in particular produces a disproportionate amount of soot (60% as much smoke as the baseline case, which involves 50 times more megatonnage) because it assumes that *all* the warheads land in the densest parts of 1000 cities. Moreover, for this scenario, the authors essentially assumed that the inner city occupied twice as large an area as in the baseline, that the net smoke emission was doubled, and that the fuel loading was two times as high as in the baseline case. Thus, this scenario should be regarded as an excursion from the baseline case.

This 100 Mt city scenario is sometimes protrayed as indicating that there might be a very low "threshold" in terms of megatonnage for the onset of nuclear winter effects. However, by the time this scenario has reached the threshold for altering the climate, it is well past the threshold for direct, human disaster: To produce so much smoke, this scenario involves the destruction of the centers of nearly all the cities in the developed countries with populations greater than 100 000.[6]

The results from TTAPS one-dimensional model are qualitatively consistent with several two- and three-dimensional climate models that appeared at about the same time. All examined the climate impact of approximately the same quantity of soot.[7-9] Most found that the continental surfaces would experience large coolings but by maximum temperature drops that were between a factor of 2 to 3 times smaller than those calculated by TTAPS—largely because they included the impact of warming by the oceans. Still these multidimensional climate models were not fully realistic. These sophisticated computer models were originally developed to simulate the Earth's climate under very modest perturbations to normal conditions. They are greatly challenged to understand the climate following the very drastic changes introduced by the dust and smoke clouds after a nuclear war.

In the several years since the possibility of a nuclear winter was first announced, these climate models have been continuously modified to make the simulations increasingly more realistic. As a result, the models now predict considerably smaller—although still significant—drops in the temperatures for the northern hemisphere following injections of large quantities of smoke and dust. The impact is predicted to be much stronger if the injection of the smoke and dust occurs in the summer months.

Uncertainties in the amounts of smoke

To make any quantitative calculations of the impact of dust and soot, the TTAPS group had to bring together estimates of many uncertain parameters. For many of these parameters, there are few measurements to guide the estimates, and often the existing measurements do not correspond to the conditions that might prevail in the event of a nuclear attack. Even the choice of a baseline scenario requires a judgement about the size and nature of a potential nuclear conflict. The baseline scenarios in the TTAPS, NAS, and other studies are admittedly large-scale wars, but they still involve only half the accumulated strategic arsenals.

The predicted quantity of soot is the product of many uncertain parameters. One must begin with estimates of the area that might be ignited by the thermal pulse of an atomic weapon. For forests, the fire start probability is strongly dependent on season, moisture level, terrain, and many other factors. For cities, it depends on the building height, density, construction materials, window area, etc. Next, one must estimate the amount of fuel available to be burned, the fraction actually consumed in the fires, and the fraction of the fuel that is converted to smoke particles. An additional parameter is how much smoke might be removed by condensation within the fire plume, with the ashes coming down in a "black rain," such as that which fell over Hiroshima. Table II indicates the range of parameter values as estimated by one analysis.[10] That table shows that the predicted amount of smoke might vary

Table II. Smoke quantities for ranges of parameter values (from Ref. 10).

Parameter type	Values			Total
	Forest	Outer city	Inner city	
Area burned (1000 km²)	100–250	76–228	4–12	180–490
Fuel loading (g/cm²)	0.1–0.5	1–4	4–20	
Emission of submicron smoke particles (g/g)	0.01–0.04	0.01–0.04	0.01–0.04	
Fraction left after immediate scavenging	0.5–1	0.5–1	0.5–1	
Total smoke (Tg)[a]	0.5–33	1–61	4–240	6–334
Specific absorption³ (m²/g)	1–3	2–6	2–6	
Hemispheric average Absorption optical depth[a]	0–0.1		0.1–4.8	0.1–4.9

[a] The ranges shown for total smoke and optical depth are the mean plus or minus two standard deviations. To estimate the mean and standard deviation, we assumed that the parameters entering the calculations for total smoke and optical depth have equal probability of assuming any of the values between the ranges shown.

over two orders of magnitude. Some of these uncertainties may be resolved with further experimentation, but others may remain wide.

Nuclear winter calculations require not only estimates of the quantity of dust and smoke that might enter the atmosphere, but also estimates of how these particles affect the transmission of sunlight. (See, for example, Ref. 11.) The dust particles scatter but do not appreciably absorb the incoming sunlight, and much of that scattered light is still forward directed. By contrast, the smoke particles absorb strongly as well as scatter light in the visible wavelength region. Both types of particles are largely transparent to infrared radiation.

The solar attenuation produced by a collection of atmospheric particles is often represented by a parameter called the optical depth τ. It determines the attenuation of a beam of light I_0 along a path at a fixed zenith angle θ, according to the relationship

$$I = I_0 \exp(-\tau/\cos\theta),$$

where I_0 is the initial solar intensity and I is the intensity reaching Earth's surface. For optical depths equal to or greater than one, the solar attenuation becomes significant. The equation above indicates only the quantity of sunlight removed from the direct beam. If the particles scatter light in the forward direction, the above equation will tend to overestimate the actual solar attenuation, because the formula does not account for the light scattered away from the direct beam that nevertheless reaches Earth's surface.

The optical depth depends both on the density of the particle layer in the atmosphere and on the optical properties of the particles themselves. The optical properties can be summarized in terms of a parameter called the specific cross section, which tells the effective area (in m²) of the incoming beam that will be scattered for every gram of material present. Materials will be

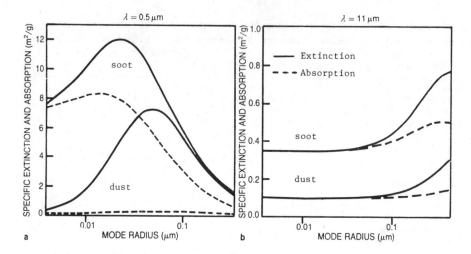

Figure 2. Specific extinction (solid curves) and absorption (dashed curves) for dust and soot particles at both (a) visible wavelengths (0.5 μm) and (b) infrared wavelengths (11 μm). The mode radius is the radius of the particles approximately midrange in size for a collection of particles of differing sizes.

characterized by both a specific *scattering* cross section and a specific *absorption* cross section ψ. The sum of the scattering and absorption cross sections is called the specific cross section for *extinction*. The specific cross section can be combined with an estimate of the surface density σ of particles to calculate the optical depth as follows:

$$\tau_i = \psi_i \sigma ,$$

where i denotes scattering, or absorption, or extinction. Sample values of specific absorption and extinction are shown in Fig. 2 (Ref. 12) for dust and soot particles at both visible and infrared wavelengths. Each curve represents the specific cross section for a collection of particles with a log normal distribution, plotted as a function of the mode radius for that distribution. Thus, for example, the solid curve in Fig. 2(a) indicates that the specific absorption for a collection of soot particles centered around 0.1 μm in radius (a radius typically assumed in nuclear winter scenarios) is about 3 m²/g. The specific absorption is nearly zero for the dust particles, indicating the reason why their impact on temperatures is much smaller.

Figure 2 indicates that the optical properties of smoke particles vary strongly with size. The optical properties also depend on the amount of elemental carbon present within the smoke particles. For example, smoke from well ventilated fires tends to have a higher fraction of elemental carbon (which is responsible for the strong absorption) than does smoke from smoldering flames. Smoke from forest fires is typically much more white (less absorbing) than smoke from the burning of petroleum products. As a result of all these variable factors, Table II indicates that the specific absorption of

Implications of nuclear winter

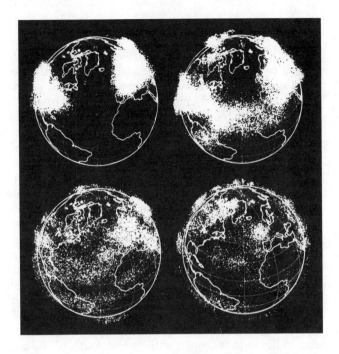

Figure 3. Dot plots illustrate the global distribution of 150 Tg of smoke 1, 5, 10, and 20 days (going clockwise from upper left) after injection of smoke at five locations in the U. S., Europe, and Asia. Each dot represents about 4000 tonnes of smoke spread throughout a volume of about 450 by 450 by 3 km. From Lawrence Livermore Laboratory

smoke might vary from 1 to 3 m^2/g for forest fires and from 2 to 6 m^2/g for urban fires.

Estimates of the ranges of possible values of specific absorption listed in Table II are used there to calculate the range of optical depths that might result from the product of many uncertain parameters—assuming that the smoke is distributed uniformly over the northern hemisphere. That range of optical depths corresponds to a fractional solar transmission from 1 (no absorption) to nearly 0 (almost extinction).

Uncertainties in the climate impacts

So far we have examined just the input to the climate models. The models themselves introduce additional uncertainties. Some of the first attempts had to make simplifying assumptions. One simulation constrained the smoke from spreading while another allowed it to spread but by normal rather than by altered circulations. None included feedback mechanisms, such as changes in precipitation patterns caused by the smoke, which might in turn prolong the residence time in the atmosphere. Some results are beginning to emerge from more interactive models, but those working on such models stress the complexity of the problem and the (fortunate) absence of real experience against which to test their models in this new regime.

The following questions are some of those being addressed by the current models:

(1) *Will the patchy distribution of smoke mitigate its impact?* The question of patchiness has been examined by MacCracken and Walton.[13] They have simulated the spread of smoke from four sources in the U.S., Europe, and Asia. Some of their results are summarized in Fig. 3, which shows the distribution of 150 Tg of smoke from these five sources as a function of time after injection in July. Each dot represents about 4000 tonnes of smoke spread throughout a volume of 550 000 km^3. In these simulations, the authors have related the scavenging rate of the aerosols to the removal rate for water vapor in the atmosphere. They point out the need for more accurate algorithms to represent the scavenging process. The problem is particularly difficult, the authors write, because actual precipitation occurs on scales that are much smaller than the scales of a few hundred kilometers that are treated by the climate model. Their results suggest that, although patchiness is significant at first, within a few weeks the smoke distribution from four discrete sources is not much different from that from an initially uniform distribution of smoke.

(2) *Will the smoke spread to the southern hemisphere?* The results from MacCracken and Walton[12] indicate that some smoke does spread to the southern hemisphere, producing only modest cooling there. Two other studies[14,15] also indicate that smoke will spread south of the equator. In these simulations, the temperatures drop in the midlatitudes of the southern hemisphere, but not below freezing. In the simulation by Malone and his colleagues, the resultant cooling in the southern hemisphere is significant only if the war occurs in the summer and not in the winter.

(3) *How fast will the smoke be removed from the atmosphere?* Most of the earliest simulations assumed that smoke was removed at rates typical of the unperturbed climate. As the models are made increasingly interactive, they are attempting to simulate the impact of the perturbed atmosphere on the particle removal rates. Malone *et al.*[14] have uncovered perhaps the most important factor determining the residence time of soot particles in the atmosphere: As the soot absorbs solar radiation, it becomes heated, expands, and some portion of it gets lofted far into the stratosphere, the layer of the atmosphere between about 13 km (at midlatitudes) and 50 km, from which micron-sized particles are only slowly removed. Thus the lifetime of the smoke in the atmosphere may be prolonged by a factor of 10 (in the summer simulation) over what it would be if the smoke were not interactive at all. Figure 4 illustrates that the smoke has risen to high altitudes and has moved towards the equator 20 days after a July injection of 170 Tg of smoke between initial heights of 2 to 5 km. Figure 4 also shows that a nonabsorbing tracer has remained at much lower altitudes and at higher latitudes. Thompson's model shows qualitatively similar lofting effects.

(4) *How dependent is the nuclear winter effect on season?* The models are indicating that nuclear winter would be far more severe if war occurred in the late spring or summer, when the solar insolation is most intense. Thompson finds that only 1% of the smoke spread south of the equator in January compared to 11% in July. Malone *et al.*[14] report much greater impact of smoke lofting in July than in January, when the solar heating is reduced.

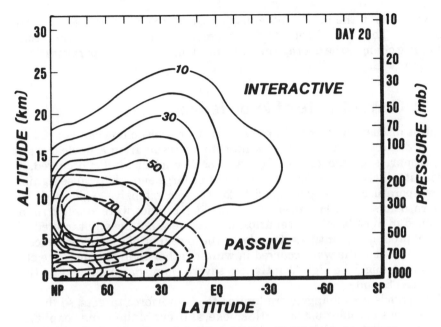

Figure 4. Smoke that absorbs sunlight (labeled "interactive") is heated and rises to very high altitudes compared to a passive tracer that does not absorb at all (labeled "passive"). Contours indicate that mixing ratio of smoke to air, in units of 10^{-9}. Curves show distributions 20 days after the injection of 170 Tg of smoke between heights of 2 to 5 km, as simulated in a three-dimensional global atmospheric circulation model. Because smoke has been lofted above the level where precipitation occurs, ten times as much smoke as tracer remains at this point. (From Ref. 14.)

(5) *How sensitive are the results to the initial distribution of smoke?* Malone *et al.* find not only that the smoke particles are removed more quickly in the January simulation, but also that the results are more sensitive to initial heights of injection of smoke in January than in July. In particular, they studied cases in which the particles were initially distributed between 2 and 5 km and compared them to cases where the smoke particles were uniformly spread from 0 to 9 km. The residence times differ by less than 20% for the two profiles in July but are more than a factor of 3 apart for the January simulation.

(6) *How sensitive are the results to the quantities of smoke?* MacCracken and Walton[13] calculated that some regions of Asia might experience maximum cooling by as much as 35 °C when the amount of smoke injected was 150 Tg. By contrast, they find virtually no significant temperature changes, and possibly even a little warming, when that quantity of smoke is reduced by 10%, to 15 Tg. Similarly, Malone *et al.*[14] report that temperatures might drop as much as 25 °C in small regions of deep interiors of the northern hemisphere when 170 Tg is injected during July. However, if only 20 Tg were injected in July, their model predicts temperature reductions of only 4–6 °C during the

first two weeks in the northern hemisphere, and no effects in the southern hemisphere. Thompson[15] reports that decreasing the smoke injection from 180 to 60 Tg changes the average midlatitude, land-surface temperature drop from 23 to 14 °C.

Current state of knowledge

From this discussion of uncertainties, it is clear that no one can yet either prove or disprove that a nuclear winter might result from a nuclear war. Many scenarios are conceivable. If a large-scale nuclear conflict occurred, it seems quite likely that optical depths might be large enough to cause significant temperature drops (perhaps 5 to 20 °C) over extensive regions of the northern hemisphere. However, if the nuclear conflict were smaller than envisioned in most baseline scenarios, if cities were not heavily targeted, if the event produced quantities of smoke on the lower end of the current uncertainty range, or if the war occurred in winter rather than summer, the climate impact might be significantly less severe than that predicted by the initial TTAPS study.

If the nuclear exchange were limited to counterforce targets, so that it generated considerable dust but little smoke, as in one of the scenarios of Fig. 1, the optical depths might still be about one. According to a recent analysis,[16] an optical depth of about one was caused by the very large eruption of the Tambora volcano in Indonesia in 1815. In the subsequent year, the average annual temperature in the northern hemisphere was estimated to be less than 1 °C below normal,[16] but the weather patterns were very erratic. Because of such unusual events as June frosts in New England, that time period has been described as "The Year Without Summer."[17] This historic event then indicates that nuclear winter might cause not only changes in the average climate but erratic and unpredictable weather patterns as well.

Implications for civil defense

Because the nuclear winter predictions remain quantitatively uncertain, it may be premature to undertake civil-defense preparations but it is not too soon to understand what measures might be desirable should the predictions prove valid. The major impact would be on agriculture, which would already be stressed by other effects of a nuclear war. (See Chap. 10.)

A worldwide committee of scientists, called the Scientific Committee on Problems of the Environment (SCOPE) organized by the International Council of Scientific Unions, has recently completed a study of both the physical aspects of major climatic effects of nuclear war, such as those discussed above, and the ecological and agricultural effects.[18] The latter panel recognized that "agricultural systems are very sensitive to climatic and societal disturbances occurring on regional to global scales, with reductions in or even total loss of crop yields possible in response to many of the potential stresses." The report concludes that the majority of the world's population is at risk of starvation in the aftermath of a nuclear war.

The greatest concern over nuclear winter is that it might occur in the spring or summer, at the peak of the growing season, threatening the agricultural yields for that year. The SCOPE study enumerated a number of different factors that would affect agricultural productivity: "...insufficient thermal time for crops over the growing season; shortening of the growing season by reduction in frost-free period in response to average temperature reductions; increasing of the time required for crop maturation in response to reduced temperature; the combination of the latter two factors to result in insufficient time for crops to mature prior to onset of killing cold temperatures; insufficient precipitation for crop yields to remain at high levels; and the occurrence of brief episodic events of chilling or freezing temperatures at critical times during the growing season."

Pimental[19] points out that most of the U.S. crops are tropical plants, which produce a high yield during the relatively short three to four month summer season. These plants require daytime temperatures of 20 to 30 °C for growth and fruit formations. Lower temperatures are expected to decrease yields. If the growing season temperatures were reduced by just 1 °C, the corn yield might fall by 7%. Some crops such as cabbage and winter wheat can tolerate subzero temperatures, but only if they are gradually acclimated to them. Hence, a sudden freeze brought on by the onset of nuclear winter conditions might kill even these colder-weather crops.

Reducing the light level on individual leaves by 50% usually does not affect the growth and yield of crops, because plants are accustomed to growing with many of their leaves all or partially shaded. However, reducing the overall sunlight level by only 10% can decrease the yield.[20] Thus a 99% or even 50% reduction in sunlight possible under nuclear winter conditions might prevent crop production.[19] A 50% reduction in rainfall might also hinder normal crop production in most agricultural regions.[19]

In addition to these problems directly created by nuclear winter, agriculture would face other problems in the wake of a nuclear war. As mentioned in Chap. 10, even without a nuclear winter, crop production could be decreased 40%–60% by lack of fertilizer and herbicides, and 20%–40% by lack of hybrid seeds. If the war destroyed access to petroleum resources, manual labor might have to replace the extensive mechanization. Whereas, in the U.S. today, only about 10 hours of farm labor are required to raise one hectare (about 2.47 acres) of corn, the same task would require 1200 hours of manpower without mechanization.[19] If the food could somehow be produced, society would still have to solve the problem of distributing it in the face of a greatly disrupted transportation system.

Long-range recovery thus requires advanced planning for mobilization of labor forces, as well as stockpiling of fuel, fertilizer, pesticides, and seeds. The SCOPE-ENUWAR study found that very few nations had stockpiled sufficient food stores to feed their populations for many months, even if those stores could be uniformly distributed to the citizens. Civil-defense planners might want to ensure that greater food stores are stockpiled. The stockpiles might include seeds for planting after a possible nuclear war that can survive in weather colder than might be typical of a particular region. It may be especially important to select crops that are rather tolerant of large temperature fluctuations.

The effects of nuclear winter would certainly extend beyond the borders of the nations involved in the nuclear conflict. Not only would these noncombatants fail to receive the agricultural products, and many other goods and services normally exported by these nations, but many nations, especially those in the northern hemisphere, would probably face the same climate disruptions.

If the nuclear war occurred in the winter, its impact would probably be less severe than if it occurred in the spring or summer. Most climate models indicate that the average temperatures drops would be lower in the winter, and in any case, most plants are dormant at this time. However, survivors already struggling to rebuild a greatly damaged society would be still further stressed—both physically and psychologically—by the cold and darkness. They might also experience violent storms or other unpredictable weather shifts that could result from the large perturbations to the normal atmosphere.

Conclusion

If the worst forecasts of nuclear winter were to prove true, this effect would be sufficiently severe and extensive to cast a great cloud of pessimism over plans for civil defense. It might dash hopes of some that large portions of the population might survive relatively unscathed from the effects of nuclear weapons. There may be few places in the northern hemisphere where one might hope to escape for long from its impact. The strongest statement one can presently make about nuclear winter, however, is that it is highly uncertain. Some climate impact can probably be expected, but how and when it occurs may make worlds of difference.

References

1. R. P. Turco, O. B. Toon, T. P. Ackerman, J. B. Pollack, and C. Sagan, Science **222**, 1283 (1983).
2. P. J. Crutzen and J. W. Birks, in Ambio (Journal of the Royal Swedish Academy of Sciences) **11**, 114 (1982). Reprinted in *The Aftermath* (Pantheon, New York, 1983).
3. *Long-Term Worldwide Effects of Multiple Nuclear Weapons Detonations* (National Academy Press, Washington, DC, 1975).
4. *The Effects on the Atmosphere of a Major Nuclear Exchange* (National Academy Press, Washington, DC, 1985).
5. Phys. Today (No. 3), 17 (1984).
6. *Concise Report on the World Population Situation in 1979* (United Nations, New York, 1980), Department of International Economic and Social Affairs, Report No. ST/ESA/SER.A/72, Population Studies, No. 72.
7. V. V. Aleksandrov and G. L. Stenchikov, "On the Modeling of the Climatic Consequences of Nuclear War," U.S.S.R. Academy of Sciences Computing Centre Report, 1983.

8. C. Covey, S. H. Schneider, and S. L. Thompson, Nature **308**, 21 (1984).
9. M. C. MacCracken, in the Third International Conference on Nuclear War, Erice, Sicily, August 1983, Lawrence Livermore Laboratory Report No. UCRL-89770, 1983.
10. B. G. Levi, and T. Rothman, "Nuclear Winter: A Review of the Key Factors," Center for Energy and Environmental Studies, Princeton University, Report No. PU/CEES 197, 1986.
11. A. A. Broyles, Am. J. Phys. **53**, 323 (1985).
12. V. Ramaswamy and J. T. Kiehl, J. Geophys. Res. **90** (1985).
13. M. C. MacCracken and J. J. Walton, in Proceedings of the International Seminar on Nuclear War, 4th Session, 19–24 August, 1984, UCRL 91446, 1984.
14. R. C. Malone, L. H. Auer, G. A. Glatzmaier, M. C. Wood, and O. B. Toon, Science **230**, 317 (1985).
15. S. L. Thompson, Nature **317**, 35 (1985).
16. R. B. Stothers, Science **224**, 1191 (1984).
17. H. Stommel and E. Stommel, Sci. Am., June 1979.
18. *Environmental Consequences of Nuclear War, SCOPE 28* (Wiley, Chichester and New York, 1985), Vols. I and II.
19. D. Pimental, Cornell University Report No. 85-1, 1985.
20. O. Bjorkman, *Encyclopedia of Plant Physiology 12A, Physiological Plant Ecology I. Responses to the Physical Environment,* edited by O. L. Lange, P. S. Nobel, C. B. Osmond, and H. Ziegler (Springer-Verlag, Berlin, 1981), pp. 57–107.

Chapter 10
Long-range recovery from nuclear war

Ruth H. Howes and Robert Ehrlich

The long-range consequences of a nuclear war are difficult to predict as they depend on the scenario assumed for the conduct of the war and on the effects of multiple nuclear explosions. Nevertheless, planning for survival following a nuclear exchange must consider them. A nuclear war will affect agriculture, industry, and the structure of the government. It may also perturb the complex ecosystem in which we live. Examination of the probable impacts of a nuclear exchange in each of these areas leads to the overwhelming conclusion that a major nuclear exchange should be prevented. There are some measures which might be taken as part of a civil defense program which would serve to mitigate several long-range effects of nuclear war.

Introduction

Successful civil defense in a nuclear war aims for not only the immediate survival of the population after a nuclear attack but also for continued survival in the event of the possibly severe ecological consequences of a nuclear exchange, such as nuclear winter. Civil defense must also provide means for a national recovery of the economy, industry, government, and for the continuation of agriculture. The degree to which a nuclear attack will disrupt agriculture, industry, the economy, and the national government clearly depends on the scenario of the attack, and there are many possible attack scenarios.

Furthermore, the long-range consequences of a large nuclear exchange are difficult to predict in detail. For example, phenomena like nuclear winter, with probable global cooling caused by smoke and dust in the atmosphere, are studied using complex computer models of atmospheric behavior, and suffer from the uncertainties inherent in such modeling techniques and their input parameters. Relatively well-understood effects such as global radioactive fallout interact with complex natural ecosystems which are themselves not understood completely. Thus discussion of long-term recovery from a nuclear attack also involves considerable intrinsic uncertainty. Nevertheless, any serious discussion of civil defense must consider possible long-term consequences of the use of nuclear weapons. Immediate survival of a substantial portion of the population is of little significance if the survivors die within two years from starvation, or if the government and society which survive ignore the basic human values and rights for which this nation stands.

Since food is essential to survival and because the topic has been studied, restoration of agriculture after a nuclear attack is considered first. The continuation of industry and the economy has been the subject of less research and is treated second. Finally, the continuity of the social structure of the nation is briefly discussed. The uncertainties in all these discussions must be stressed. Predicted consequences of a nuclear exchange depend not only on the scenario assumed for an attack and the complex calculations of its consequences to the environment. These uncertainties also depend on the responses of citizens and social systems to the stresses of a nuclear war, and these responses can only be inferred from studies of historical disasters with different time scales and different causes.

Restoration of agriculture

The production of food crops and livestock will be interrupted by three mechanisms. First, a certain percentage of crops will be destroyed by the immediate effects of nuclear explosions and the resulting radioactive fallout. (Note: Chap. 4 discusses the effects of nuclear weapons.) Second, multiple nuclear explosions may disrupt the Earth's climate both through the injection of dust and smoke into the atmosphere and by the depletion of the ozone layer. Third, modern agriculture in the United States depends heavily on the use of chemical fertilizers, pesticides and herbicides as well as petroleum-powered farm machinery. In addition to this reliance on the chemical industry, the national transportation system and the national petroleum industry are essential to modern agriculture.

The effects of nuclear radiation on crops, livestock, and forests were extensively studied during the 1950s and 1960s. The tolerance of plants to radiation depends on the species of plant involved. Crops such as rice are relatively tolerant to large doses of radiation while barley is damaged by doses a factor of 5 smaller. To reduce crop yields by 50% requires gamma doses between 1 and 20 krad, which doses would occur over only a small fraction of the agricultural land in the United States. There is evidence that plants with a greater number of chromosome pairs per unit volume of cell nucleus are more vulnerable to radiation damage than genetically simpler species. The destructive effects of radiation doses depend on the rate at which the radiation is received. A rapidly delivered dose does more harm than an equivalent dose delivered more slowly. Plants respond relatively slowly to radiation, so observations of damage often require several weeks or months. The response of a plant to a given dose of radiation depends sensitively on the stage of the plant's development at the time of the exposure. Recovery from damage is obviously influenced by such environmental conditions as temperature and moisture.[1]

Most early radiation damage studies used gamma radiation as the source of radiation doses to plants. Then experiments done during nuclear tests in the atmosphere showed that fallout particles were trapped by plant foliage.

Thus crops received large doses of beta radiation which damaged them more severely than gamma radiation (several papers in Ref. 2). The exact composition of fallout particles depends on the magnitude of the explosion which produced them and the height of the burst. If the fireball of the burst touches the ground, soil is sucked into the fireball and its activated components increase fallout. Detailed doses to crops are difficult to predict since such properties of the fallout particles as solubility in water depend on the soil and bomb materials which condensed with the radioactive nuclides to form them. These same chemical and physical properties of the fallout determine the rate at which radioactive atoms are absorbed by plants and incorporated into the food chain.[3] Uncertainty in the composition of fallout complicates predictions of its biological effects. Finally, the long-lived nuclides ^{90}Sr (28.1 year half-life) and ^{137}Cs (30.0 year half-life) form part of the global fallout, become part of the structure of plants and eventually enter the food chain. They will not, however, produce doses sufficient to damage the plants themselves or cause reduction in crop yields.[4] Like plants, seeds are subject to radiation injury which can cause failure to sprout if the dose is at least of the order of magnitude of the lethal doses for mature plants. Seeds in storage containers should be sufficiently well protected to escape radiation injury.[5]

In the event of a nuclear war, crops will almost certainly receive direct damage from nuclear explosions, since the major missile bases in the United States are in agricultural regions. Fire will destroy some fields and other crops will be killed outright by radiation. The short-term survival of crops in a nuclear attack will be approximately 25% for a medium-sized nuclear exchange directed at military and major industrial targets.[6] This number, of course, depends on targeting doctrine assumed for the attack and the season of the year when the attack occurs.[6] Crops contaminated with fallout will be slightly damaged by radiation and can be safely consumed by people and livestock if the fallout particles are thoroughly washed from the plants. Civil defense must provide instructions for doing this.

Livestock is also vulnerable to death and damage from fallout radiation. Fifty percent lethal doses range from around 200 rads for large ruminants up to 900 rads for poultry.[7] For humans the 50% lethal dose is about 450 rads. Large animals suffer severe radiation burns from fallout particles which land on their backs and stay there, as well as from eating pasture contaminated with fallout. Muscle from animals who have been exposed to radiation can be used for meat if the animals have not undergone internal hemorrhaging, which spoils the meat. Contaminated fodder will ruin meat not taken from the muscle.[5] Milk may be contaminated by ^{131}I immediately after the attack and by ^{90}Sr in the long run and should be used with caution.[4]

The civil defense measures needed to protect livestock are relatively straightforward. If possible, animals should be brought into barns which will protect them at least from direct contact with fallout and from beta burns, although barns generally have low protection factors. If uncontaminated feed is available, it should be used. Several days without food will be better for animals than feeding with heavily contaminated foodstocks. Radiation doses from fallout drop by a factor of 100 in the time from one hour after the burst to two days after the burst, and then by another factor of 10 after about two weeks (see Appendix A.4). Thus danger from contaminated feed decreases rapidly. Animals will need water which is not contaminated with radioactivity, as will people. This implies an immediate need to tap sources of ground water, which may not be an easy task in the probable absence of electric power and the extreme shortage of petroleum products. In the not improbable event that a major portion of the livestock cannot be sheltered, remaining animals will have to be saved for breeding purposes. This will release stocks of stored grain for human consumption, but necessitates concern with the provision of adequate protein in the diets of survivors, particularly several months after the attack.[8]

A final effect of radiation on agriculture may be through selective destruction of different plant and animal species. Any catastrophe tends to throw an ecosystem back from its climax stage to a more primitive stage of development favoring less complex organisms. In many cases this is a well-understood and positive effect.[9] Fires are set to maintain pine forests, for example. However, in the case of severe radiation exposure, insects and other smaller organisms which are often pests to agriculture may survive better than the larger predators which normally control them.[10,11] Such pests can cause large-scale damage to agriculture unless controlled by herbicides and pesticides which are likely to be unavailable.

Although radiation will stress crops and livestock, large-scale disruptions in the climate of the Earth may prove more threatening to the continuation of agriculture. There are at least two mechanisms for this climate disruption: ozone depletion and nuclear winter. Nuclear winter was discussed in Chap. 9. The detonation of high-yield warheads produces large quantities of nitrogen oxides which may be injected into the upper atmosphere. Once in the atmosphere, they undergo a complex series of photochemical interactions which ultimately cause the breakdown of ozone molecules. This process depletes the layer of ozone in the atmosphere, which protects organisms on the Earth's surface from the damaging effects of short wavelength ultraviolet light.[12,13] Severe depletion of the ozone layer damages plants in varying degrees depending on their sensitivity to ultraviolet radiation. In an extreme case it has been speculated that exposure to ultraviolet light might blind animals and insects. Blind insects would be unable to pollinate crops thus leading to the failure of agriculture, and a disastrous effect on the human species.[14] Fortunately, it seems that such an extreme disaster is unlikely even in the largest possible nuclear exchange. The exact extent of damage to the ozone layer depends critically on the size of weapons detonated, since warheads with yields less than 500 kt do not produce fireballs large enough to inject appreciable quantities of nitrogen oxides into the ozone layer. The effect also depends on the total number of weapons exploded. The consensus of calculations of this process is that the explosion of half the world's stockpile could result in a 30% to 45% depletion of the ozone layer which would heal itself by half in two to three years and be mostly restored in five to six years (a list of studies of ozone depletion is found in Knox et al.[15]). The resulting increase in ultraviolet radiation at the Earth's surface might result in the extinction of some species of plants and animals but the ecosystem as a whole would certainly survive this perturbation.[12] The trend toward smaller warheads mounted on more accurate MIRV's (multiple independently targetable reentry vehicles) also reduces potential damage to the ozone layer since the nitrogen oxides they produce remain near the Earth's surface.[16] The NAS Study of 1985 found a 17% decrease in ozone for their baseline case. Unfortunately, oxides of nitrogen also form photochemical smog when they appear near the surface of the Earth.[17] The severity of this smog as well as the problem of other chemicals produced by the burning of cities has not been estimated quantitatively at this time.

Current evidence indicates that a major nuclear conflict will probably result in global cooling, although the magnitude of the effect is not at all certain. The mechanisms for removing smoke from the atmosphere in the long and short run, and the quantity of smoke which will be injected are both subjects of considerable speculation. Thus, at the present time, no one can draw definite conclusions about the magnitude of nuclear winter, but all evidence indicates that such an effect could occur.[13,18]

In planning the resumption of agriculture after nuclear war, the probable cooling brought on by nuclear winter must be considered. It may well destroy one season's crops or delay replanting by as much as a year depending on the time of year when the war occurs. Cold-tolerant species such as winter wheat might be prepared for a first planting after a nuclear war. Livestock will also need shelter and food for the cold period if it is to survive. In a particularly severe scenario with widespread temperature drops of 40 °C, all agriculture and indeed much of all plant life could be destroyed. In the absence of severe cooling, disruption of normal weather can severely damage crops as can darkness severe enough to interrupt or decrease photosynthesis.

Modern agriculture relies on chemical fertilizers, pesticides, herbicides, specialized seeds, and mechanized farm equipment to produce its bountiful yields. Modern farming methods for one acre require 120 gallons of fuel and four hours of labor as opposed to 500 hours of manual labor. Crop production would be decreased by 40% to 60% by lack of fertilizer and herbicides, and 20% to 40% by lack of hybrid seeds—even without nuclear winter.[19] In addition a modern agricultural field is usually a monoculture and not a natural ecosystem so that when it is disturbed, it does not tend to restore itself without human intervention.[20] If agriculture is to be restored following a nuclear attack, farmers will need supplies of petroleum fuels, fertilizers, pesticides, and herbicides. They will need seed for crops which can survive both nuclear winter and the increased ultraviolet radiation which will follow the dissipation of the smoke. Regions heavily contaminated with fallout may require power for deep-furrow plowing which would place much of the fallout underground. Maintenance of needed supplies will require the restoration of the transportation system and the petroleum and chemical industries. Petroleum stocks are among the more vulnerable elements of our society. Provision will have to be made to allocate scarce fuels to agriculture, as was done during the 1973 embargo. Priority must be given to transportation and production of vital agricultural chemicals and to transportation of harvested crops and stored food.[4] The electric power grid will probably not be functioning in the immediate aftermath of a nuclear exchange, and its absence will further hinder the resumption of modern agriculture.

In the event that an attack destroys either the national petroleum industry or severely damages the road and rail systems, agriculture will revert to methods using coal, animal, or human power when limited local petroleum and chemical supplies are exhausted. In such a situation, harvested crops could not be distributed to cities or food processing plants which had survived. Population would probably migrate spontaneously to rural areas or be moved there by government orders. Agriculture would become labor intensive and less productive so that a large percentage of the survivors would be employed in this sector of the economy and thus not available for other efforts at reconstruction. Since petroleum refineries are tempting targets and extremely vulnerable, they may well be largely destroyed.[16] Plans made in advance to supply farmers with needed fuel and chemicals will greatly speed agricultural recovery and thus free workers to restore other damage from the attack.

Recovery of industry and the economy

The industrial structure of the United States will be damaged in at least three ways during a nuclear attack. First, buildings and equipment will be destroyed. Second, many key and highly skilled personnel and managers will be killed. Third, damaged transportation and communication systems will impede the flow of raw materials and finished goods, creating bottlenecks in the system.

Experimental studies of protection of industrial buildings and machinery during a nuclear attack were conducted by the Boeing Company.[21,22] These studies demonstrated that delicate machinery could be protected from blast and fire produced by a nearby nuclear explosion, although not from a direct nuclear hit. Individual machines are packed with metal chips from milling machines or with crushed newspapers. The packed machines are then covered with dirt. This procedure was borrowed from Soviet manuals. Boeing estimates that one of its major plants, the Boeing Auburn fabrication facility, could be prepared in this manner in three or four working days by its regular staff who could pack the key one-third of the equipment and move the earth necessary to bury it.[22] More time is needed to dig out the equipment and reconnect it for the start of manufacturing. The source of power to run the equipment is not clear either. This process will obviously require the services of skilled laborers who must be persuaded to work on the industrial recovery rather than on their personal recovery. The Boeing studies conclude that saving buildings is not as important as saving machines since the machines can be set up in temporary shelters, as demonstrated by the Soviets and the Germans during World War II and by the North Vietnamese in the Vietnam War. The protection of industrial equipment by these relatively simple means is designed to force an attacker to expend more weapons to destroy a given portion of the industrial capacity of the United States. The strategic importance of this aspect of civil defense is emphasized in the final Boeing study[22] and a study by Jones and Thompson.[23] These studies do not stress the fact that implementing this form of protection is expensive in its own right. Time and skill are required to set up machines for a manufacturing process. Delicate equipment will be knocked out of alignment and damaged as it is packaged for burial. Any mistake in lubrication and sealing will expose it to direct contact with the dirt and sand in which it is buried. A false alarm in which these civil defense measures were implemented would be extremely costly, and in itself a staggering blow to the economy. The value of the protection of key equipment in a nuclear attack must be balanced against the costs of implementing the protection, the conceivable danger of provoking an attack by starting such a process, and the costs of a false alarm. The situation is similar to the case for and against population evacuation, and many of the same arguments are applicable.

Protection of skilled workers may be accomplished by evacuation or sheltering as discussed in earlier chapters. If manufacturing operations are to resume quickly, it is clearly desirable that workers be sheltered near the equipment they operate. Thus extensive shelters for skilled workers and their families will be needed on or near the sites of plants. If workers are to be evacuated, alternative reporting centers will need to be established. A major effort would need to be made to educate workers to report to the correct places after the attack. More than one worker would need to be trained to perform critical operations in the manufacturing process, and backup supervisors will be needed. This educational effort would be expensive in terms of man-hours dedicated to training, but it is an essential element of industrial civil defense.[24–26]

Under existing federal civil defense plans, each individual corporation is expected to make plans to reestablish its productivity and its corporate structure under the law. Debts and other financial and legal obligations should be recorded and honored. Individual plants within large corporations will be responsible for resuming their own production.[24,27] The restoration of electric power and water supplies will be key elements in industrial recovery. Workers in these areas will require special protection and organization.[28] The degree to which transportation and communication systems will be disrupted depends on the size and structure of the nuclear attack sustained by the nation. Studies by Katz[29] have examined various levels of industrial destruction produced by a set of differing attack scenarios. Katz points out that it may be difficult to assess even the extent of industrial damage sustained in a nuclear attack. Selective destruction of critical industries may be designed specifically to bottleneck the entire economy. Since skilled workers and industrial facilities are concentrated in a relatively few urban areas, a small nuclear attack on major cities could destroy more than half of our industrial capacity. According to Katz and studies cited in his work, industrial recovery to 80% of the gross national product (GNP) achieved before the attack will require from five to ten years after the attack under the assumption that the war stops and a stable government organizes the recovery. Other authors (e.g., Laulan[30]) feel that the disorganization of the economic structure would be so severe as to destroy the possibility of economic recovery for decades if indeed it could ever be achieved. The complex social and economic infrastructure of a modern society is certain to be heavily damaged by even a light nuclear exchange.[31]

Many observers believe that it would take more than a decade for the best designed and implemented program for industrial civil defense to restore productivity so that the GNP represents a substantial fraction of its prewar level. This assumes that agriculture is successfully restarted so that people can be spared from food production to organize industrial recovery. A large nuclear war will severely damage most if not all the industrial nations in the northern hemisphere so there is little hope of outside aid reaching the struggling society. Competition for scarce resources will certainly be acute in the years immediately following a nuclear attack. Unless a central authority directs their allocation, society may tend toward a set of isolated and primitive city states as described in Appendix C of the study by the OTA.[16] Industrial and economic recovery thus depend on the existence of a central national government and sufficient social structure to organize a complex interconnected economy. National systems of communication and transportation will have to be restored before a large scale industrial recovery can begin.

Recovery of national government and social order

No data exist on the reactions of nations to a large scale nuclear war. Authors base their thinking either on extrapolation of the effects of strategic bombing during World War II or on the massive death and destruction produced in medieval Europe by the Black Death. These models do not fit the short time scale and the magnitude of the death and destruction predicted for a modern nuclear war. The only certainty seems to be that the nation will face problems on an unprecedented scale.[16,29,31–33]

Many national, community, military, and business leaders will be killed in the attack. Elaborate and classified preparations exist for the preservation of the military chain of command so there will probably be a surviving commander in chief. The surviving population will be isolated in relatively small sheltered groups. Very high radiation levels from fallout over perhaps 10% of the country (depending on whether or not population centers are targeted) will necessitate shelter stays of at least two weeks, and emergence from shelters for short periods of time for weeks after that. Even in areas where radiation levels are not lethal, survivors will probably remain in shelters out of fear of arriving radiation. Commercial radio and television stations will probably not operate during this period either because they lack electric power or because critical components have been burned out by electromagnetic pulse during the attack. In many regions, survivors will know nothing of the status of the national government nor whether there are other survivors.

National resources will be destroyed to an unprecedented level. In particular, organized medical care will be terribly difficult to provide since many hospitals are in industrial areas and will probably be destroyed in the attack. Trained medical personnel will also be killed in large numbers. Many survivors who are wounded in the attack and who do not die immediately will require months of intensive medical care. The financial and legal systems will be in shambles since fragmented groups of survivors will exist as isolated communities. Lastly, the national government may well be engaged in the continuation of the war which produced the nuclear attack in the first place. Under some scenarios, the commander in chief will be in the process of ordering nuclear counterattacks or conventional military operations. The United States may experience a series of continuing nuclear or conventional attacks.

Studies based on a hypothetical nuclear attack on Boston estimate that each uninjured physician will need to treat more than 1700 injured survivors. If the physician spends 15 minutes with each patient and works 16-hour days, this will require more than two weeks before each patient sees a doctor once.[34] Many of these patients will be burn victims and require a quality of medical care that will not be available. Even transportation of patients is likely to prove difficult in the immediate aftermath of a nuclear attack. Within crowded shelters, survivors may suffer from the effects of exposure to radiation, heat, lack of water, and poor air quality. Sanitary facilities may be severely overcrowded particularly in the probable absence of running water. The symptoms of radiation sickness (diarrhea and nausea) are similar to those produced by stress so many people may think their condition is worse than it actually is.[35] Sublethal doses of radiation decrease the body's resistance to disease. Under such conditions, respiratory and other common diseases may spread rapidly among survivors.[36] Drugs for treatment and even morphine for the injured are likely to be unavailable unless shelters are stocked with large supplies of them. This is difficult since many drugs have shelf lives on the order of a year.

Restoration of national government and social order must begin with ending the war and establishing the crippled United States as a national entity in the eyes of the rest of the world. Locating leaders of other friendly nations, let alone those of hostile nations, may be difficult in the confusion following nuclear attacks. It will be extremely difficult to establish social order until on-going nuclear attacks stop. Conventional military invasions from either the original nuclear adversary or third parties not involved in the nuclear war may attempt to seize territory and partition the United States. The first responsibility of a restored national government will be to end hostilities with the United States intact as a nation, and to see that wars do not reignite.

The national authorities must then communicate with groups of survivors and establish a central government of the nation with authority over local officials. Survivors must feel a sense of national unity and authority so that resources from lightly damaged regions can be shared with the population of harder hit areas. In a time of poor transportation and communication, local governments must assume the actual operation of their communities, and will tend to regard themselves as the only legitimate political entity. The difficulties involved in accomplishing these tasks are obviously enormous.

Once established, the national government must organize the distribution of the scarce resources available for reconstruction. This will require the imposition of martial law to prevent looting and stealing and to enforce rationing. Even in smaller disasters such as tornadoes, the National Guard frequently serves as a temporary government in the stricken area. Survivors must be set to work caring for the wounded and disposing of the dead. Work crews will be needed to decontaminate areas and to restore essential sanitary services. The population will need food, shelter, and water, particularly if they face an extended period of extreme cold. People must be persuaded to abandon work on their personal property to assist in the restoration of transportation, communication, and other national needs such as electric power, coal, and petroleum. To accomplish these things, a directed economy will be necessary for the first years following the attack. If severe ecological damage such as a nuclear winter has been sustained, the tight controls of the recovery period will have to be extended if there are to be any survivors at all. Initially high levels of radiation will make work difficult at least in some areas. Workers will need careful radiation monitoring and organization into shifts if they are to avoid radiation sickness and/or damage. The presence of corpses, the insects which feed on them and the bacteria which infest them, in conjunction with poor sanitary conditions and the absence of medical care, drugs, and vaccines will hold a high potential for the rapid spread of diseases such as typhoid, plague, and tuberculosis. In some areas, rabid animals may present a major problem.[37] A rudimentary public health system must be established to control epidemics.

If the strictly ordered society can survive the two or three years after a nuclear attack as a national entity, then the central government must turn its attention to recreating a society based on a free economy and the human rights we value. To establish a free economy there must be confidence in a monetary system. Initially, survivors will tend to barter since one cannot eat or wear money. If a complex economy is to be sustained, a monetary system with a banking structure for the availability of capital is essential since barter becomes too complex for all but communities at a minimum subsistence level. Planning before a war can help in this area of restoration by saving legal and financial records and by providing a supply of money for exchange, but no preplanning can restore people's confidence in that money.

The government must cope with the burden of surviving but incapacitated wounded and with survivors suffering from a variety of severe psychological disorders.[29,32] Families must be reunited and provisions made for the care of the large number of orphans. An educational system must be established. A substantial portion of the surviving population will receive large doses of radiation which will not cause immediate illness after the attack, but which may later lead to the development of cancer or genetic defects in their offspring.[16,38] The severity of these effects is difficult to calculate because of the uncertainty in the type of nuclear attack involved and the effects of low-level radiation. If, for example, nuclear power reactors and waste storage sites are targeted, fallout will contain longer-lived radioactive isotopes and dose rates will be higher.[39] Evidence from studies of the survivors of Hiroshima and Nagasaki indicates an initial increase in the rate of leukemia peaking about eight years after the nuclear detonation and appearing first in people who were children at the time of the attack. By 20 years after the attack, the risk of cancer for persons exposed to 100 rads or more was about 2.5 times that of unexposed persons. Different types of tumors appear at various times after radiation exposure and have been correlated with exposure in varying degrees of certainty. There is no evidence of increased cancer in children born to exposed parents although children irradiated *in utero* were subject to malformations and mental retardation, and some stillbirths occurred.[40] Such effects are difficult to document in detail because of lack of information on the radiation dose received by each individual. The genetic effects of radiation on succeeding generations are even more difficult to assess. While it is known that such effects occur in animals, human data are as yet inconclusive. In our current society these medical effects of radiation would be considered large and unacceptable. In the society following a nuclear war, they will probably represent only a small additional burden on facilities charged with the care of those crippled by the short-term effects of the attack and its immediate aftermath. Civil defense education measures such as training people to read radiation meters can greatly reduce average radiation exposure to the population, assuming, of course, that meters are distributed before the attack.

Lastly, the national and regional governments must abandon military control of society and reestablish a democratic government with careful guarantees of basic human rights.

Conclusions

In 1965, Eugene P. Odum summarized a conference on the ecological effects of nuclear war by pointing out two great uncertainties in predicting its long-range consequences: interactions among the various effects of nuclear weapons and the unknown effects of the very massive use of weapons whose individual effects are well studied.[41] His conclusions are valid today.

It is possible that a nuclear winter or other environmental disaster produced by a nuclear exchange would be so severe as to render all civil defense futile or alternatively that the environmental consequences will be mild compared to the immediate destruction of a nuclear exchange. Civil defense cannot afford the luxury of ignoring possible long-term effects of nuclear war because they are difficult to calculate.

Bibliography

Nuclear winter:
  (1) National Research Council, *The Effects on the Atmosphere of a Major Nuclear Exchange* (National Academy Press, Washington, DC, 1985). An up-to-date and thorough review of the subject.
  (2) R. P. Turco, O. B. Toon, T. P. Ackerman, J. B. Pollack, and Carl Sagan, "Nuclear Winter: Global Consequences of Multiple Nuclear Explosions," Science **222**, 1283 (1983). The paper that started it all and which is still the basis of many discussions.

On long-term effects:
  (1) Jeannie Peterson (editor), *The Aftermath* (Pantheon, New York, 1983). This issue of *Ambio* covers a good broad range of subjects.
  (2) Arthur M. Katz, *Life After Nuclear War* (Ballinger, Cambridge, MA, 1982). This is an excellent review of the impact of nuclear war on a modern society.

Discussion questions

(1) How complete is our understanding of the long-range effects of a large nuclear war? Can we justify civil defense expenditures to counter these effects in the face of the uncertainties which exist concerning them?
(2) How might appropriate civil defense measures reduce the long-range effects of nuclear war on the United States? What would these civil defense measures be?
(3) Do you feel that plans to reestablish our government after a nuclear war are realistic? Why or why not?
(4) Discuss the effects of the black plague in medieval Europe and the strategic bombing of World War II as models for the effects of nuclear war.
(5) What lessons learned at Hiroshima/Nagasaki are applicable to a comprehensive civil defense program?
(6) Why does not what happened at Hiroshima/Nagasaki extrapolate to a full scale nuclear war?
(7) What are the implications of "nuclear winter" for civil defense planning at the present time?

References

1. A. H. Sparrow, S. S. Schwammer, and P. J. Bottino, in Brookhaven National Laboratory Report No. CONF-700909, proceedings of a symposium 15–18 September, 1970 (BNL, Upton, NY, 1971), p. 670.
2. D. W. Bensen and A. H. Sparrow, in Ref. 1.
3. C. F. Miller, in Ref. 1, p. 81.
4. R. S. Russell, B. O. Bartlett, and R. S. Bruce, in Ref. 1, p. 548.
5. S. Glasstone and P. J. Dolan, *The Effects of Nuclear Weapons*, 3rd ed. (U. S. Department of Defense and the Energy Research and Development Administration, Washington, DC, 1977).
6. S. L. Brown, in Ref. 1, p. 595.
7. M. C. Bell, in Ref. 1, p. 656.
8. R. J. Sullivan, K. Guth, W. H. Thoms, and F. L. Adelman, *Survival During the First Year After a Nuclear Attack* (System Planning Corporation, 1979), Report No. SPC 488.
9. T. T. Kozlowski and C. E. Ahlgren, *Fire and Ecosystems* (Academic, New York, 1974).
10. G. M. Woodwell and A. H. Sparrow, in *Ecological Effects of Nuclear War*, edited by G. M. Woodwell (BNL, Upton, NY, 1965), USAEC Report No. BNL-917, p. 20.
11. G. M. Woodwell, in *The Aftermath*, edited by J. Peterson (Pantheon, New York, 1983).
12. National Academy of Sciences, *Long-Term Worldwide Effects of Multiple Nuclear Weapons Detonations* (National Academy Press, Washington, DC, 1975).
13. Committee on the Atmospheric Effects of Nuclear Explosions, Commission on Physical Sciences, Mathematics and Resources, National Research Council, National Academy of Sciences, *The Effects on the Atmosphere of a Major*

Nuclear Exchange (National Academy Press, Washington, DC 1985), also known as NAS 1985.
14. J. Schell, *The Fate of the Earth* (Knopf, New York, 1982).
15. J. B. Knox, M. C. MacCracken, M. H. Dickenson, P. M. Gresho, F. M. Luther, and R. C. Orphan, *Program Report for FY 1982 Atmospheric and Geophysical Sciences Division of the Physics Department* (Lawrence Livermore Laboratory, Berkeley, CA, 1982), Report No. UCRL-51444.
16. Office of Technology Assessment, *The Effects of Nuclear War* (Allanheld, Osmun & Co., Totowa, NJ, 1980), first issued in 1979.
17. P. J. Crutzen and J. W. Birks, in Ref. 11 [also in Ambio **11**, 114 (1982)].
18. C. W. Weinberger, "The Potential Effects of Nuclear War on the Climate," Report to the U. S. Congress, 1985.
19. D. Pimentel, *The Impact of Nuclear War on Agricultural Production* (Cornell University, Ithaca, 1985), Research Report No. 85-1.
20. V. M. Stern, in Ref. 1.
21. Boeing Aerospace Company, *Industrial Survival and Recovery from Nuclear Attack: A Report to the Joint Committee on Defense Production, U. S. Congress* (Boeing Co., 1976).
22. J. W. Russel and E. N. York, *Expedient Industrial Protection Against Nuclear Attack* (The Boeing Co., 1980).
23. T. K. Jones and W. S. Thompson, *Orbis* **22**, 681 (1978).
24. Department of Commerce, *Preparedness in the Chemical and Allied Industries* (Department of Commerce, Washington, DC, 1968).
25. Department of the Interior, *A Guide to the Defense Electric Power Administration* (Department of the Interior, Washington, DC, 1972).
26. National Petroleum Council, *Civil Defense and Emergency Planning for the Petroleum and Gas Industries* (National Petroleum Council, 1964), Vols. 1 and 2.
27. U. S. Office of Minerals and Solid Fuels, *Civil Defense in the Minerals and Solid Fuels Industries* (U. S. Department of the Interior, Washington, DC, 1964).
28. National Academy of Sciences, *Civil Defense—Project Harbor Summary Report* (National Research Council, Washington, DC, 1964), Publication No. 1237.
29. A. M. Katz, *Life After Nuclear War* (Ballinger, Cambridge, MA, 1982).
30. Y. Laulan, in Ref. 11.
31. Joint Committee on Defense Production of the Congress of the United States, Committee on Banking, Housing, and Urban Affairs, U. S. Senate, *Economic and Social Consequences of Nuclear Attacks on the United States,* (U.S. GPO, Washington, DC, 1979).
32. F. C. Ikle, *The Social Impact of Bomb Destruction* (University of Oklahoma, 1958).
33. E. I. Chazov and M. E. Vartanian, in Ref. 11.
34. H. H. Hiatt, in *The Final Epidemic*, edited by R. Adams and S. Cullen (University of Chicago, Chicago, 1981).
35. P. J. Lindop and J. Rotblat, in Ref. 34.
36. H. L. Abrams and W. E. Von Kaenal, *N. Eng. J. Med.* **305**, 1226 (1982).
37. H. L. Abrams, in Ref. 34.
38. Stockholm International Peace Research Institute, *Nuclear Radiation in Warfare* (Taylor and Frances, New York, 1981).
39. B. Ramberg, *Destruction of Nuclear Energy Facilities in War* (Lexington, Cambridge, MA, 1980).
40. S. C. Finch, in Ref. 34.
41. G. M. Woodwell, Ref. 10.

Chapter 11
Political and psychological issues in civil defense*

Robert Ehrlich and Ruth H. Howes

Introduction

Political issues concerning civil defense can loosely be grouped in four principal categories.

Short-term feasibility. Are civil defense measures, particularly crisis relocation, feasible under realistic scenarios for a nuclear war?

Long-term feasibility. Even if it provides some measure of short-term success in reducing casualties, does civil defense offer more than a "band aid" solution to the long-term hazards to survival, particularly nuclear winter?

Provocation. Does civil defense undermine or enhance deterrence? Does it make nuclear war more or less likely?

Costs and benefits to society. What are the economic and social costs of implementing a major civil defense program in peacetime?

In what follows we shall address political and psychological issues as they relate to each of these categories. Questions of physical feasibility are addressed more extensively elsewhere.

Short-term feasibility

Civil defense measures to be successful must first protect people against the immediate and short-term effects of nuclear weapons: blast, thermal radiation, fires, prompt nuclear radiation, and delayed radiation due to fallout. Protection against these short-term dangers can be achieved, in principle, through sheltering, evacuation (relocation), or a combination of relocation and sheltering. Specific protective measures can be quite effective when one is faced with specific dangers, e.g., taking cover to avoid flying glass or staying in the shadows to escape thermal radiation. This presumes, of course, that the survivors are not too close to the point of detonation and that they do not fall victim to some other unforeseen danger. Nevertheless, statistically a popula-

*Parts of this chapter have been reprinted from *Waging Nuclear Peace* by Robert Ehrlich by permission of the State University of New York Press.

tion which takes protective measures against well-defined effects of nuclear weapons is bound to suffer fewer short-term casualties than one which does not.

Crisis relocation is perhaps the most controversial of all civil defense measures, and many legitimate questions have been raised about its feasibility. The questions primarily relate to how long a warning period is necessary for relocation to be considered. There are many scenarios in which a nuclear war might start without sufficient warning to permit an evacuation. These include (1) a surprise attack (with no detectable prior Soviet evacuation), (2) a very rapid escalation from a conventional war with the U.S.S.R., (3) by accident, or (4) by design—started by third parties seeking mutual U.S.–U.S.S.R. destruction. There are obviously many such situations for which a crisis relocation would be unfeasible, just as there are also many other scenarios for which it would be feasible. Most people do not refrain from buying life insurance just because they know that their policy will not pay off under certain circumstances, e.g., death during a nuclear war. The key factor in making such judgements would seem to be how likely are those circumstances that make a plan feasible and how likely are those that make it unfeasible. That question is not an easy one to answer either for life insurance or civil defense.

Of the circumstances necessary for a crisis relocation to be successful, ample time is the most important. Various events could precipitate a directive to evacuate our cities. One of the most plausible would be the Soviet evacuation of their cities. Such a Soviet evacuation would probably be quickly detected by spy satellites (assuming they are still working) or by other means, but what would the U.S. response be? The alternatives might include (1) increasing the readiness of our nuclear forces, but not ordering an evacuation of U.S. cities (in that case, assuming the Soviet evacuation became public knowledge, the result might be mass turmoil and spontaneous evacuation of U.S. cities); (2) making up a highly complex evacuation plan, disseminating and implementing it with no prior planning; (3) launching a suicidal first strike on the U.S.S.R. before their evacuation ran its course; (4) retargeting U.S. missiles at the Soviet evacuation centers, which would be counter to stated U.S. targeting objectives and which would only work in the unlikely event that the U.S.S.R. evacuation did not lead to a great deal of dispersal; and (5) reaching the Soviet leader on the hot line in an attempt to forestall the impending catastrophe.

While these alternatives may be neither exhaustive nor exclusive, (5) would seem preferable to the others. The president could, of course, call the Soviet leader without ordering a U.S. evacuation, possibly even threatening a first strike attack unless the U.S.S.R. ordered its "citizen hostages" back to their cities. However, it is hard to imagine a Soviet leader complying with such a request when it was fear of an imminent U.S. attack that might have stimulated the order to evacuate in the first place.

Aside from having ample time, a successful evacuation of cities would also apparently require a "cooperative enemy" who would refrain from launching an attack while the evacuation is in progress. Otherwise, the casualties might be even greater than had no evacuation taken place. This point is contestable, since even if only a few hours notice were available a fraction of the evacuating population might have a much better chance of survival. The retargeting objection to evacuations seems to have several flaws. First the percentage of the U.S. land area subject to the blast and thermal effects of nuclear detonations in an all-out attack would probably be only about 5%. Fallout covers larger areas and fallout shelters would be a necessary component of any viable relocation plan. Thus a relocation which takes the form of a dispersal would leave far fewer people vulnerable to the direct effects. Moreover, it is unclear whether an enemy would want to "waste" nuclear warheads in attacking the relocated population rather than destroying the evacuated cities which contain the economic and industrial potential of the nation and which could be reoccupied. Finally, such restraint by the enemy need not result from any benign attitude, but rather from the deterrence stemming from the mutual destruction that each side could inflict on the other's fleeing evacuees.

In a very severe but prolonged crisis, perhaps involving the prior use of tactical nuclear weapons in Europe, or perhaps even following the start of Soviet evacuation of their cities, it is hard to imagine that large percentages of the U.S. urban population would resist the impulse to evacuate, particularly once a spontaneous evacuation began to get underway. At the Three Mile Island accident, for example, it is estimated that 40% of the nearby population evacuated spontaneously. In the absense of a system of public or private shelters or a viable ballistic missile defense, the only choice a U.S. government would have during an extreme crisis would be between directing an evacuation based on a hastily formulated and minimal plan, or allowing a spontaneous evacuation to occur with all the chaos that is likely to entail. Blast shelters require much less warning time than crisis relocation; however, they are much more expensive than fallout shelters. At the present time there seems little inclination on the part of either the military, federal or local governments, or most private individuals to construct them.

Long-term feasibility

Although it has often been asserted that civil defense could, in fact, have a very significant short-term impact on casualties, its impact on the long-term recovery of the society is less clear. For example, even if the Central Intelligence Agency (CIA) assessment of Soviet civil defense is correct, and, in the most favorable circumstances, the U.S.S.R. were to suffer not more than 10% casualties following a U.S. retaliatory strike, the survival of the U.S.S.R. as a functioning society would be far from assured. As the physicist Henry Kendall notes, a second U.S. retaliatory strike would have devastating consequences on the U.S.S.R. economy and social structure, wiping out the bulk of its industry, oil refineries, military and political targets, and cities.[1] Moreover, given the scarcity of material goods and, more importantly, the destruction of an economy that makes production of new goods possible, a case can be made that the eventual recovery of a society is more likely if there are fewer initial survivors.[2]

However, the morality of this specific anti-civil-defense argument is equivalent to that of taking just enough survivors into a lifeboat so that, given the limited provisions on board, all can be reasonably certain of surviving for some time, rather than taking as many as the boat can hold. Quite apart from the dubious morality of not making serious civil defense efforts that might save many tens of millions of lives since it might endanger long-term recovery, the truth of the asserted proposition is by no means self-evident. It may well be that to reconstruct a devastated society, the most important asset is not material, but rather the accumulated skills and knowledge of its surviving individual citizens. It is possible that the great bulk of our accumulated knowledge adapted to a highly technological civilization would have little relevance to the utterly unprecedented situation in which the survivors would find themselves. However, it would seem equally absurd to think that the extremely difficult and perhaps impossible road back to a functioning society would necessarily be made more difficult by the presence of too many physicians, engineers, policemen, craftsmen, farmers, and people generally.

Moreover, civil defense supporters note that the situation for survivors might not be so different from the marginal existence in many parts of the world today, or from the way everybody lived not long ago. Obviously extremely difficult, possibly insurmountable problems would exist in restoring production of food and other essential goods, as well as equally important problems in distributing these goods. These problems are so serious that civil defense planning before the fact can only make things less miserable for the survivors and for the possible eventual reconstruction of society. Such planning cannot assure that there will be a government in charge to follow through on the plans, that people would receive and obey instructions of such a government, or that society would eventually recover.

It may well be that following an all-out U.S.–U.S.S.R. nuclear exchange, the survivors would indeed envy the dead. Just as in World War II concentration camps, there would be many among the survivors who would give up all hope and would not bother to try to survive. There would also be those who would fight desperately to survive. If a nuclear war were to result in catastrophic climatic or ecological changes of the magnitude some scientists have suggested, then the long-range prospects of human survival would indeed be bleak. Thus the magnitude of the climatic effect termed "nuclear winter" is a key issue for civil defense. At the present time, calculations of long-term climatic effects are plagued by vast uncertainties (see Chap. 9). Although a catastrophic temperature decline might make civil defense efforts futile, it is also possible that a temperature decline would be of less than catastrophic proportions, and that would make certain civil defense measures potentially very important, e.g., having stored food reserves. Moreover, to the extent that the governments of the United States and the Soviet Union believe predictions of climatic catastrophe, they may be moved to actions that mitigate that possibility. Such actions might include a further move to smaller weapons, a change in targeting strategy to avoid cities, the construction of effective active defenses against nuclear delivery systems, particularly ballistic missiles, or the mutual reduction in the size of arsenals. Each of these measures would increase the effectiveness of civil defense in promoting long-term recovery because the scope of the devastation would be less.

Finally, even if we grant there is uncertainty over whether a civil defense program would help the long-term recovery following an all-out nuclear war, there are other events for which a civil defense program would be of unquestionable benefit. Such eventualities include (1) a limited nuclear war with the U.S.S.R., (2) a limited nuclear war initiated by a third party, (3) an accidentally fired U.S. missile hitting a U.S. target, (4) a nuclear terrorist attack, (5) a catastrophic nuclear reactor accident, and (6) other non-nuclear disasters. While in many of these cases there is unlikely to be much warning time before the event, much less warning would be needed. For example, suppose a nuclear war were initiated by a third party or by an accidental launch of a small number of Soviet missiles. Whether such actions would quickly lead to all-out war is impossible to say. While many untargeted cities hundreds of miles from the detonation would have zero warning time before the blast itself, they would have ample warning of perhaps a day (depending on wind velocity) of the approaching radioactive fallout cloud. In the event that the strategic defense initiative is successful in constructing active defenses against nuclear missiles, crisis relocation might save lives following detonation of those warheads which leak through the defenses. Such measures would limit the damage produced by these smaller nuclear events and greatly speed the long-term recovery of the society.

Provocation

Our current strategic policy is that of mutual deterrence, which relies on the threat of nuclear retaliation by one side to prevent the first use of nuclear weapons by the other. Civil defense opponents argue that a sizable U.S. civil defense effort would be destabilizing because it would prompt the U.S.S.R. to add still more weapons to its nuclear arsenal in order to ensure the capability to destroy the United States. However, there is no indication that the United States has found it necessary to build a larger arsenal in order to keep pace with the Soviet civil defense effort. Deterrence does not depend on one's ability to kill in retaliation exactly as many of the enemy as were killed in the initial attack. The point is rather that if there is a great disparity in that regard then this factor, along with others, e.g., the sizable advantage to the side that strikes the first blow, could significantly undermine deterrence and increase the probability of nuclear war. Thus asking whether an expanded U.S. civil defense effort would stabilize or destabilize the U.S.–U.S.S.R. strategic balance is, in effect, asking whether the action would increase or decrease the probability of nuclear war.

Perhaps the most persuasive argument that has been advanced against civil defense is that it would foster the belief that nuclear war might be survivable or winnable. While a nuclear war might be survivable in the technical sense of there being survivors, recovery in the affected countries would by no means be assured. An expanded civil defense effort would therefore need to be accompanied by a realistic portrayal of how terrible a nuclear war would really be. Unfortunately, past governmental public-education efforts have failed to show the consequences of a nuclear war and the extraordinary hardships survivors would undergo. However, ordinary citizens in the U.S.S.R. seem no more eager than U.S. citizens to have a nuclear war, despite their government's sizable civil defense effort. Therefore, it is hard to say why a U.S. civil defense effort accompanied by extensive public education would make nuclear war less abhorrent to U.S. citizens. In fact it might well become more abhorrent and thereby strengthen deterrence. Thus instead of taking the position that we should work on methods to reduce the risk of nuclear war rather than methods to survive one (civil defense), we may alternatively take the position that we should simultaneously promote civil defense and other methods to reduce the risk of war.

Much more important than dollar expenditures on civil defense is how the money is to be spent. If a U.S. civil defense program is going to accomplish anything, it is essential that past mistakes be corrected, particularly in producing educational information for the public that shows the real dimensions of a nuclear catastrophe. It is difficult for the public to be optimistic that officials charged with the mission of creating a civil defense plan really know what they are doing when one reads media accounts of administrative procedures for forwarding mail and collecting taxes following a nuclear war. (This credibility problem is due to a lack of sophistication on the part of both administration spokesmen and the media. Moreover, the problem is compounded by politicians who want to make civil defense appear as absurd as possible in order to kill the program.)

An argument against a U.S. evacuation plan is that a U.S. evacuation of its cities even after a Soviet evacuation took place might well trigger a Soviet first strike. A government that orders an evacuation of its cities is taking an extreme and provocative act, one likely to be interpreted by the other side as a preparation for a preemptive strike. An evacuation therefore invites a preemptive first strike while the evacuation is underway. In this context, an effective civil defense program can be viewed as preparation for nuclear war fighting. On the other hand, mutual deterrence depends on a rough balance between the capabilities of the sides. This deterrence would seem more assured if both sides are perceived as being capable of inflicting comparable damage on one another. If it is perceived that an effective Soviet evacuation might allow them to suffer only 10% casualties (the estimate in a 1978 CIA study), while the U.S. might suffer as much as 75% fatalities with no evacuation or warning, there could conceivably be a significant U.S. temptation to strike while the Soviet evacuation was in progress. Similarly, a Soviet military planner might well conclude that a brief "window of opportunity" had opened in which a preemptive first strike would inflict very much greater damage on the United States than the Soviet Union would absorb in retaliation. According to that reasoning, the Soviet military might be tempted to be much more aggressive given the temporary U.S. vulnerability.

By itself, an effective civil defense program can hardly be viewed as an inducement to adventurism, given the devastation a nuclear war would inflict even with civil defense. Ship captains do not navigate more recklessly because their ship is equipped with lifeboats. However, it must be remembered that civil defense must be evaluated as a part of our overall strategic policy.

One indication of how the Soviets would view a greatly expanded U.S. civil defense program is supplied by the Soviet booklet *Whence the Threat to*

*Peace?*³ The first edition of this booklet was prepared by the Soviet Defense Ministry in response to the controversial 1981 U.S. Department of Defense booklet, *Soviet Military Power*. The earlier U.S. booklet described Soviet military might in stark terms and it included a section on Soviet civil defense as well.⁴ The Soviet booklet was more "balanced" in that it compared the two superpowers' military arsenals and then proceeded to attack all the elements of the proposed U.S. buildup under President Reagan as undermining an alleged existing nuclear parity.³ The one notable omission in the Soviet attack on the U.S. military buildup was any reference whatsoever to the proposed significant expansion of U.S. civil defense efforts. One may infer from this omission that the Soviets have found no fault with a U.S. program designed to save lives and that they do not view it as provocative or destabilizing. An alternative explanation of their silence on this one point might be that they consider the program so unimportant as to be merely a waste of U.S. dollars. This possibility must be viewed in the context of the Soviets' own large expenditures on civil defense.

It is even possible that the Soviets view U.S. inattention to civil defense in an ominous light. A Soviet defense planner might well regard a near-total U.S. neglect of civil defense as a strong indication that the U.S. option of choice is a first strike or that we in the United States really have little appreciation for what war would be like on our own soil. Lacking such an appreciaton, we might therefore react in panic in a military confrontation and thereby blunder into a full-scale nuclear war. Whether or not this ominous connotation may seem far-fetched, Soviet civilians and officials have voiced it in individual conversations.⁵ The interpretation given to actions or inactions of one country by people in another country is not easy to second guess, particularly when two countries harbor such deep suspicions of each other's intentions. Whatever the Soviet view of an expanded U.S. civil defense effort, Soviet officials seem eager to downplay the scope of their own effort.⁶ Since there are many in the U.S. who equate civil defense with preparation for a nuclear war, the Soviets would not wish to be seen as preparing for war, particularly when denouncing a U.S. administration on that score.

Finally, both the Soviets and the United States must evaluate civil defense in view of changing strategies. President Reagan has stated that the policy of the United States is to initiate a vigorous research program in an attempt to develop defenses against ballistic missiles which will make nuclear weapons "impotent and obsolete." If the program initiated by the president to develop ballistic missile defenses, known as the Strategic Defense Initiative (SDI), is successful, it will prevent the vast majority of Soviet warheads from reaching the United States. Because the Soviet Union also would deliver nuclear warheads with bombers and cruise missiles launched from the ground, ships, and airplanes, SDI will have to be accompanied by a defense against these systems if it is to protect the U.S. from nuclear attack. No mechanical system ever works perfectly, so even under the condition that an extremely effective defensive system is deployed around the U.S., a few nuclear weapons may be expected to be detonated over targets in the U.S. In this case, a program of fallout shelters, at least in areas near likely targets such as missile silos, and in downwind areas, could reduce the impact of a nuclear exchange on the U.S. The Soviets appear to view SDI as essentially an offensive effort and would probably view civil defense associated with it as threatening deterrent stability. It is clear that the current major reevaluation of our strategic policy may well necessitate a reevaluation of civil defense as part of that policy.

Costs and benefits to society

A massive civil-defense effort could be relatively expensive although when it is considered in the context of such weapons systems as the Trident submarine, it seems less so. Historically the United States has spent far less on civil defense than have the Soviet Union or Switzerland (see Appendix A.9 and Chaps. 3 and 8). Federal and state dollars are at a premium and civil defense has not been politically popular in this country even during our current military buildup. The present policy of combining civil defense planning with planning for natural disasters like floods and tornados represents one economy move in civil defense planning, but like all economy moves the plans may fall far short of what is needed.

A crisis relocation plan would necessitate great care in implementation. A mass evacuation of U.S. cities, particularly if it were accompanied by measures such as burying industrial machines, would impose an enormous cost to the society. Industry and commerce would grind to a complete halt during the evacuation and would take time to restart. Some injuries and illness will result even in the most orderly evacuation and these will be much greater in the event of a major panic. The fear of an impending nuclear attack and the massive disruption of everyday life brought about by the evacuation will tend to aggravate psychological disorders. Patients evacuated from hospitals and nursing homes will suffer physical discomfort and possible worsening of their conditions. The legal tangles over loss of pay by evacuees who did not work or damage to buildings at the host area will be formidable in their own right. In the event the evacuation is a false alarm, it will thus be expensive. Consequences include damage to unattended homes and industries, loss of production, restart costs, social disruption, and the fact that the government had "cried wolf" in ordering the evacuation in the first place, so that people might be less responsive to a second order to evacuate. Nevertheless, these costs, considerable as they may be, need to be weighed against the millions of lives to be saved in a nuclear war if evacuation can be successfully carried out.

Aside from its obvious monetary cost, civil defense carries political costs to those who advocate it. Many cities have refused to develop crisis relocation plans and the plans that have been proposed have proved extremely unpopular in designated relocation areas. Some groups of physicians have refused to participate in developing plans for medical care in the aftermath of nuclear war on the grounds that their participation harms humankind by making a nuclear war that much more likely. Other groups advocate paramilitary survival procedures. Political pressures from groups which take extreme positions in favor of or opposed to civil defense have tended to reduce civil defense to an article of ridicule or faith rather than an element of overall strategic planning. Although polls show that the general public is reasonably sympathetic to the idea of civil defense, particularly when they are informed about the large Soviet civil defense effort and lack of any comparable U.S. effort, civil defense is not an issue of great saliency to most people, who do not care enough to try to influence government policy on it. Civil defense policy has been subject to pressure from small groups of vociferous advocates of particular positions.

Finally, the implementation of civil defense carries immediate social costs. People must give up time for training and drills. They must accept the discipline necessary to organize a national civil defense program. Scarce resources must be stockpiled and will not be available for industrial use, possibly driving up the prices citizens pay for finished goods. Popular realization of the consequences of nuclear war will be strengthened by a civil defense program. The general public will have to become more convinced that a nuclear war may occur if they are to be asked to participate actively in civil defense training and drills. The psychological impact of this heightened perception of nuclear war is a matter of speculation but must be considered before a civil defense program is implemented.

Bibliography

Most of the works on the politics of civil defense are written by people who are either advocates of stronger programs or opponents of any civil defense program at all. The two documents below are overviews which seem reasonably balanced and make a good starting point for reading in this area.

(1) Gary K. Reynolds, "Civil Defense", a major issue brief from the Congressional Research Service of the Library of Congress, 1984, order No. IB84128. An excellent summary of politics in historical context which contains a very strong bibliography listing hearings and other documents.
(2) U.S. Arms Control and Disarmament Agency, *An Analysis of Civil Defense in Nuclear War* (U.S. GPO, Washington, DC, 1978), a review which unfortunately precedes SDI.

Partisan sources include the following.

(1) *Front Line*, P.O. Box 1793, Santa Fe, NM 87504. A newsletter critical of civil defense.
(2) *Journal of Civil Defense of the American Civil Defense Association*, P.O. Box 910, Starke, FL 32091. Strongly favors civil defense.
(3) "Civil Defense in the 80s: Key Aspects of the Debate on Civil Defense," issued by Physicians for Social Responsibility, 639 Massachusetts Avenue, Cambridge MA 02139. Critical of aspects of civil defense.

Discussion questions

(1) What accounts for the negative reactions some people have towards the idea of civil defense?
(2) What are the arguments for and against an expanded U.S. civil defense effort, particularly a crisis relocation plan?
(3) Could civil defense make a nuclear war more survivable in the short term? In the long term?
(4) In what ways would an expanded civil defense effort make nuclear war more likely? Less likely?
(5) If you do not think civil defense is a good idea, what would you think an appropriate U.S. response would be to a Soviet evacuation of their cities? Is that something worth thinking about? Explain.
(6) To what extent do you think that government plans to expand civil defense are directed towards saving lives rather than towards matching the Soviet civil defense effort?
(7) Does the size of the Soviet civil defense effort worry you? Explain.
(8) What are the moral and ethical questions concerned with expanding the U.S. government's civil defense program? With not expanding the program?
(9) What effect would ecological-environmental catastrophes such as nuclear winter or ozone depletion have on civil defense efforts?
(10) Would a civil defense program on a large scale make a big difference, or only a marginal difference, in the impact of a nuclear war on civil society?
(11) What impact would various kinds of civil defense measures have on peacetime diplomacy or crisis stability?
(12) What civil defense measures would be appropriate if nuclear war were considered likely in the next few years?
(13) What kind and size of civil defense program might be worth the money it would cost?

References

1. H. Kendall, *Bull. At. Sci.* **35**, 32 (1979).
2. H. Kendall, "Testimony on Civil Defense," prepared in connection with hearings before the U.S. Senate Committee on Banking, Housing, and Urban Affairs, 8 January 1979.
3. U.S.S.R. Ministry of Defense, *Whence the Threat to Peace?* (Soviet Defense Ministry, Moscow, 1984).
4. U.S. Department of Defense, *Soviet Military Power* (U.S. Department of Defense, Washington, DC, 1984).
5. L. Hecht, private communication on 23 November 1982.
6. *Oregonian*, 13 October 1982.

Appendix A

compiled by John Dowling

Appendix A.1: Lifeboat analogy to civil defense

The following piece is from an internal document circulated at FEMA, "The Chipman Report." It is an interesting example of how far analogies can go. It is included in this report for the reader's amusement. John Hassard, author of Chap. 7, has the last word.

Arguments against civil defense and a rebuttal

Some of the arguments made against civil defense were parodied as follows in a piece in the *Harvard Crimson* in 1962.

"*Recommendations by the Committee for a Sane Navigational Policy*: It has been brought to our attention that certain elements among the passengers and crew favor the installation of lifeboats on this ship. These elements have advanced the excuse that such action would save lives in the event of a maritime disaster such as the ship striking an iceberg. Although we share their concern, we remain unalterably opposed to any consideration of their course of action for the following reasons.

(1) This program would lull you into a false sense of security.
(2) It would cause undue alarm and destroy your desire to continue your voyage in this ship.
(3) It demonstrates a lack of faith in our Captain.
(4) The apparent security which lifeboats offer will make our navigators reckless.
(5) These proposals will distract our attention from more important things, e.g., building unsinkable ships. They may even lead our builders to false economies and the building of ships which are actually unsafe.
(6) In the event of being struck by an iceberg (we will never strike first) the lifeboats would certainly sink along with the ship.
(7) If they do not sink, you will only be saved for a worse fate, inevitable death on the open sea.
(8) If you should be washed ashore on a desert island, you could not adapt to the hostile environment and would surely die of exposure.
(9) If you should be rescued by a passing vessel, you would spend a life of remorse mourning your lost loved ones.
(10) The panic caused by a collision with an iceberg would destroy all semblance of civilized human behavior. We shudder at the prospect of one man shooting another for the possession of a lifeboat.

(11) Such a catastrophe is too horrible to contemplate. Anyone who does contemplate it obviously advocates it...

In conclusion, there is perhaps a parallel between an evacuation capability for our cities and lifeboats on an ocean liner. The mere fact that the lifeboats exist will not lead the ship's captain to take additional risks with his ship in bad weather or treacherous seas. Nor will the lifeboats prevent the ship from being totally destroyed by collision with a reef, iceberg, or other vessel. Nor will the lifeboats necessarily insure even the immediate survival of all the people on board. Nor will they necessarily insure the prolonged survival of those who do successfully abandon the ship in them: the lifeboats themselves may be swamped; supplies may give out before the survivors are rescued; exposure, injury, and exhaustion may take their toll. Nonetheless, the lifeboats do offer the prospect that, in the event the ship sinks, at least some people will survive for somewhile and that with luck a good number may survive to sail again in another vessel. In any event, no one would want to cross the oceans in a ship without lifeboats. In parallel fashion, simple prudence dictates that the United States should not attempt to cross the uncertain and troubled waters of the 1980s without the capability to evacuate its urban population in the event of catastrophe."

And from John Hassard, Chap. 7: "Crisis relocation plan proponents often draw analogies between having evacuation plans and wearing a seat belt in a car, or even more often to having lifeboats on a ship (this analogy lending itself nicely to visions of the Ship of State crossing the turbulent waters of the 1980s). They point out that having lifeboats does not make the captain take risks with icebergs. Opponents might find this analogy amusing, but probably also a trivialized simplification of rather complex issues. It contains the suggestion that nuclear confrontations are out of our control, like icebergs in the night. ... Ship enthusiasts might ask themselves what would be the point of lifeboats that took 'one to two weeks' to launch and then did not float. I would hope that they would do their best to warn other passengers, and to pressure the skipper to seek safe routes. They may even choose a new captain."

Appendix A.2: High-risk areas map

The high-risk areas map (Fig. 1) is from p. 9 of a 1984 GAO Report (see Chap. 3, Ref. 1). In the map FEMA identifies some 400 areas in the United States considered to be at high risk from the direct weapons effects of a large-scale nuclear attack because of proximity to important military and urban-industrial areas.

Figure 1. High-risk area map.

Appendix A.3: Standard radiation decay model for fallout

The decrease of radiation with time is discussed in Ref. 1 of Chap. 6 and also in C. M. Haaland, C. V. Chester, and E. P. Wigner, "Survival of the Relocated Population of the U.S. After a Nuclear Attack," Oak Ridge National Laboratory Report No. 5041, 1976, pp. 39 and 42. The empirical relation below holds *after* all the fallout is on the ground:

$$R = R_0 t^{-1.2},$$

or

$$t = (R_0/R)^{0.833}$$

and relates R, the intensity of radiation in roentgens/hour at time t in hours after the attack, to R_0, the unit time reference dose rate. Table I correlates the reduction factor with time.

Table I. Reduction factors for radiation intensity due to decay. (Here $R_0 = 1$, one hour after the attack.) This is the so-called 7–10 rule; after an approximate sevenfold time lapse, the radiation decreases by a factor of 10.

Time after attack		Reduction factor
(Hours)	(Days)	
1	0.042	1
6.8	0.28	0.1
46.4	1.9	0.01
316	13.2	0.001
2154	89.8	0.0001

Appendix A.4: The "penalty" table

Haaland, Chester, and Wigner (p. 66) discuss an alternate guideline to radiation exposure which they call the "penalty" table (Table II). This table gives an idea of when medical care will be needed with respect to the accumulated radiation dose.

Table II. Medical care needed with respect to radiation dose

		Accumulated radiation exposures (roentgens) in any period of		
		a	b	c
	Medical care will be needed by	One week	One month	Four months
A	None	150	200	300
B	Some (5% may die)	250	350	500
C	Most (50% may die)	450	600	...

Appendix A.5: Area dose-rates after a hypothetical attack

In order to give the reader an idea of the dose rates after a nuclear attack we provide the following table (Table III) from Haaland, Chester, and Wigner, p. 54. Here the hypothetical attack is one where 6559 megatons, of which 5951 are ground bursts, is assumed to strike industrial and population targets in the U.S. This is a rather large attack and is obviously one not limited to military targets. (Editor's note: In the U.S. the average yearly dose for people is about 0.2 R/year. This is equivalent to 0.000 022 8 R/h. About half of this dose, or 0.000 01 R/h is attributable to background radiation, the rest is due to what we ingest, medical x rays, watching TV, air travel, etc.)

Table III. Area dose-rates one year after a hypothetical attack

Dose rate (roentgens/h)	Area (sq. mi.)	Cumulative area (sq. mi.)	Percent of area of coterminous U.S.
0.1	108	108	0.004
0.01	347 000	347 000	11.7
0.001	1 089 000	1 436 000	48.5
0.000 1	944 000	2 380 000	80.3
0.000 01	274 000	2 654 000	89.5

Appendix A.6: Guidelines for shelter and operational activities

Table IV is adapted from Haaland, Chester, and Wigner, p. 63 which they took from *Radiological Defense Planning and Operations Guide*.

Table IV. Guidelines for shelter and operational activities

If outside Dose rate has fallen to: (in roentgens/h)	Activities that may be tolerated
Less than 0.5	No special precautions necessary except to keep fallout particles from contaminating people and to sleep in the shelter.
0.5–2	Up to a few hours/day outdoor activity for essentials: fires, policing, rescue, repair, getting food, water, medicine, important communications, waste disposal, exercise, and fresh air. All other activities done in shelter.
2–10	Less than an hour/day outside for the most essential purposes. Outdoor tasks rotated to minimize doses. Children outside limited to 15 minutes max/day.
10–100	Outside activities limited to a few minutes devoted to nonpostponeable tasks. People stay sheltered no matter how uncomfortable.
Greater than 100	Outside activities very risky, particularly if dose rate much greater than 100. Only leave if risk of death from fire, collapse of shelter, or thirst, or if present shelter is greatly inadequate and if better shelter is only a few minutes away.

Appendix A.7: Exposures in shelters of low protection factors

Table V is from Haaland, Chester, and Wigner, p. 77. It gives an idea of what radiation exposure is likely to occur in shelters with protection factors (PF) varying from 5 to 20 for periods of time from 4 days to 4 weeks. (Note: If you have a PF of 10, the radiation outside the shelter is then roughly 10 times what it is inside the shelter.)

Table V. Exposures in shelters of low PF. $R_0 = 2500$ roentgens/h. Fallout arrival time 13.5 hours after detonation.

PF	Exposure (roentgens) Time in shelter, beginning when fallout arrives					Comments
	4 days	1 week	2 weeks	3 weeks	4 weeks	
5	508	602	710	770	808	100% lethal
10	224	301	355	385	404	30–40% lethal
15	169	201	237	263	269	100 % radiation sickness
20	127	151	178	193	202	No medical attention required

Appendix A.8: Fallout conditions from a random assumed attack

Figure 2 is from p. 24 of a 1984 GAO Report (see Chap. 3, Ref. 1). It depicts the fallout conditions from a random assumed attack against a wide range of targets: military, industrial, and population.

SURVIVAL ACTIONS
- No shelter required under this wind condition
- Up to 2 days shelter occupancy
- 2 days to 1 week shelter occupancy
- 1 week to 2 weeks shelter occupancy followed by decontamination in exceptional areas

Figure 2. Fallout conditions from a random assumed attack against a wide range of targets on a spring day.

Appendix A.9: Costs relative to civil defense

Table VI shows per capita military and civil-defense expenditures and some comparable figures compiled by Evans M. Harrell, School of Mathematics, Georgia Institute of Technology, Atlanta, GA 30332. These numbers are accurate to within several percent.

Table VI. Per capita military and civil-defense expenditures.[a]

Current annual expenditures in the United States	Per capita
Arms Control and Disarmament Agency	$0.07
Federal civil defense	0.75
Strategic Defense Initiative (Star Wars) research	6.00
Ballistic missile defense	24.00
Orders of nuclear arms and systems	130.00
Total Federal military R&D (1986 proposed)	179.00
Department of Defense (1986 proposed)	1350.00
Some possible alternatives	
French civil defense	0.15
Swiss civil defense	33.00
National blast shelter program[b]	500.00
Strategic Defense Initiative (Star Wars) construction	860.00
	to 4300.00
Some comparable figures	
White House office	0.09
One C-5B airplane	0.53
Smithsonian Institution	0.81
Military research at M.I.T.	1.05
High energy physics operating funds	2.35
National Science Foundation	6.90
One Trident submarine	8.60
Estimate of superconducting supercollider[b]	12.90
National Aeronautic and Space Administration	32.00
Department of Energy	33.00
Military contracts of Rockwell International	36.00
Total Federal research and development	220.00
Federal deficit	950.00
Total Federal outlays	4100.00

[a] Sources: *Atlanta Constitution, Defense Monitor,* FEMA, *Physics Today, Science, Wall Street Journal, World Almanac.*
[b] These figures would be spread over several years.

Appendix B: Should we protect ourselves from nuclear weapon effects? *

Carsten M. Haaland

Simple buried-pole shelters of Soviet design have been found to withstand 100 psi overpressure in American tests. The radius R in miles at which the overpressure is P in psi from a ground burst of yield Y in megatons, is given by

$$[R = 6Y^{1/3}/[P^{0.5} - 0.02(P^{0.5} - 4)^2]]$$

for $0.5 < P < 500$. In those few areas in the U.S. where conditions might support a firestorm, safe shelters could be constructed based on lessons learned from the Hamburg firestorm of 1943 in which 85% of the 280 000 people within the firestorm survived. If the people at Hiroshima had been in simple buried-pole shelters, not one person need have perished from weapon effects even at ground zero where the overpressure was only 40 psi because the weapon was airburst to increase blast effects. Soviet propaganda for exterior consumption attempts to persuade our people that civil defense is not possible, whereas their internal line stresses that "there is not and can never be a weapon from which there is no defense." Despite the fact that individuals can be protected from nuclear weapon effects, the gloominess of a nationwide picture of the U.S. after a large attack indicates that a multilayer missile defense is required in addition to civil defense.

For over 20 years I have had the fear that there might be a sudden unexpected escalation of events between the United States and the Soviet Union that would lead to a nuclear war, despite the deterrence provided by the awesome arsenals of each superpower. Consequently, I have devoted my career to the study of possible ways of protection against the effects of nuclear weapons, with the hope and expectation that an effective civil defense would improve deterrence and thus, in a passive manner, reduce the possibility of the ultimate horror of nuclear war. In the event that deterrence should fail, these measures of civil defense could possibly save millions of lives, as I shall discuss in my presentation.

---

*Research jointly sponsored by FEMA and U.S. Department of Energy under Contract No. W-7405-eng-26 with Union Carbide Corporation. This paper was presented at a Forum Symposium on Civil Defense at the March 1984 Meeting of The American Physical Society in Detroit, MI by Carsten M. Haaland, Engineering Physics and Mathematics Division, P.O. Box X, Oak Ridge National Laboratory, Oak Ridge, TN 37830.

NUCLEAR WEAPON EFFECTS
(GROUND & LOW AIR BURSTS)
PRIMARY SEQUENCE

1. **INITIAL NUCLEAR RADIATION (INR)**
2. **INDUCED RADIOACTIVITY**
3. # THERMAL RADIATION
4. **ELECTROMAGNETIC PULSE (EMP)**
5. # SHOCK & BLAST-WIND

Figure 1. Primary sequence of nuclear weapon effects.

My talk is divided into three parts. The first part will briefly review the effects of and defenses against individual nuclear weapons. The second part will discuss possible effects of and defenses against the detonation of thousands of nuclear weapons in a war between the superpowers. The third part will present an overview of Soviet civil defense.

The three main prompt effects of the detonation of a nuclear weapon, those taking place within the first minute or so after detonation, are initial nuclear radiation, thermal radiation, and blast (Fig. 1). I will not discuss induced radioactivity or EMP because these effects are of lesser significance. Initial nuclear radiation consists of a pulse of neutron and gamma radiation that is gone in less than a second after detonation, followed by fission gammas and some delayed neutrons that may continue to have a harmful effect at the Earth's surface for about one minute after detonation. Thermal radiation at a distance is limited to those photons from the fireball that have wavelengths that are transmittable in air, that is, those in the visible spectrum, with some radiations in the near infrared and ultraviolet. The characteristics of thermal radiation are fairly similar to those of radiation from a blackbody at a temperature of 6000 to 7000 degrees Kelvin. The blast wave is a wall of compressed air that moves out from the fireball at very high pressures and velocities several times the speed of sound, accompanied by violent winds. As the

Figure 2. Frontispiece of *The Effects of Nuclear Weapons* (Available from U.S. Government Printing Office, Washington, DC 20402; price: $17.)

The Effects of Nuclear Weapons

Compiled and edited by
Samuel Glasstone *and* Philip J. Dolan

Third Edition

Prepared and published by the
UNITED STATES DEPARTMENT OF DEFENSE
and the
UNITED STATES DEPARTMENT OF ENERGY

1977

blast wave proceeds outward, the energy in the shock front dissipates and the pressure and velocity reduce to levels of sound signals.

In addition to these prompt effects, damaging and hazardous delayed effects can be produced by large-scale fires and by contamination of large areas by radioactive fallout. These delayed effects do not occur categorically. Obviously, there will be no large fires resulting from nuclear bursts over military targets, on water, in deserts, or in rocky areas where there are little or no combustible materials. Less obviously, there will be no contamination by radioactive fallout from detonations where the fireball doesn't touch the ground, unless a rainout occurs.

A third category of effects includes several possible long-range global consequences of large-scale nuclear attacks, such as depletion of the ozone layer or reduction of surface temperature due to dust and smoke in the atmosphere. I will not dwell on these effects because, if further research indicates that these effects may be serious, it appears to be within the capability of each superpower to modify their nuclear arsenals and targeting doctrines so that strategic objectives can be accomplished without causing these effects.

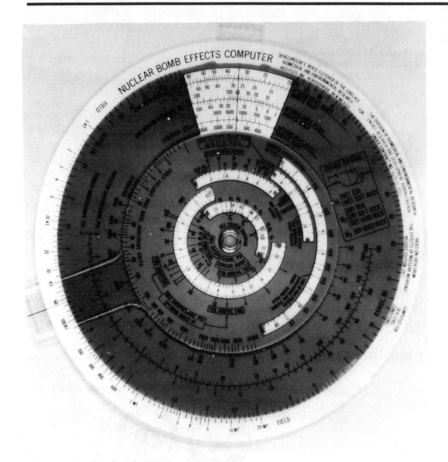

Figure 3. Nuclear bomb effects computer.

Nearly all the information presented here has been prepared from data taken from *The Effects of Nuclear Weapons*[1] (Fig. 2). This 650-page book comes with a plastic circular slide rule (Fig. 3), the Nuclear bomb effects computer, which contains a great deal of information on blast effects and crater sizes on one side, and data (Fig. 4) on initial nuclear radiation and thermal radiation on the other side.

Let us review the sequence of events following the airburst of a one-megaton, thermonuclear warhead in which 50% of the energy is produced by fusion and 50% is by fission. The height of the burst will be chosen so the range is maximized at which the blast wave on the ground has an overpressure of 10 lb per square inch, or 69 kP. This overpressure assures destruction of most unprotected industrial facilities. For a 1 Mt weapon, this burst height is 7200 ft. The Hiroshima weapon of 12.5 kt was detonated at an altitude of 1670 ft.[2] That burst height also results in a maximum range for the 10-psi overpressure on the ground from that weapon, hence, by blast scaling laws, the overpressure on the ground directly below the 1 Mt air burst will be the same as for the Hiroshima burst, namely, about 40 psi.

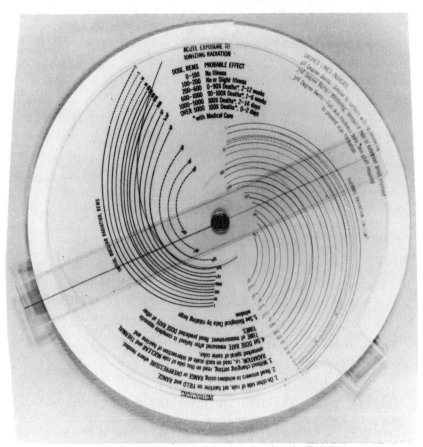

Figure 4. Reverse side of nuclear bomb effects computer.

Suppose we are standing at the edge of a cliff that rises several thousand feet above a vast plain on which lies a large hypothetical American city called Metropolis. The 1 Mt weapon is detonated over the center of Metropolis, located about 20 miles away from our position (Fig. 5). The fireball expands to almost a mile in diameter in one second, and reaches maximum thermal radiance in nine-tenths of a second. The blast wave does not reach the ground until about 2.7 sec after the burst (Fig. 6, top).

At three seconds after the burst, the heat radiating from the fireball, indicated by the white circle in Fig. 6 (top), is five times less than it was at the first second, and about 60% of the total thermal energy has been radiated. The part of the energy released only as thermal radiation from this burst is equivalent to the total energy released by the detonation of 350 kt of TNT. Figure 7 shows a typical thermal radiation history from nuclear detonations. The first flash radiates only about 1% of the thermal energy. For a 1 Mt weapon, the peak radiance in the second pulse occurs at about 0.9 seconds after detonation.

At the time of three seconds after the burst, if the atmosphere is clear, a

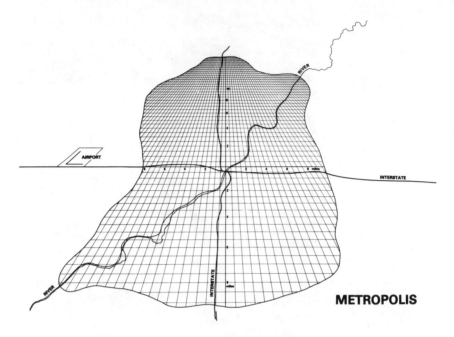

Figure 5. Detonation area of Metropolis.

dry, light-weight, khaki shirt in direct line-of-sight to the fireball will ignite within a range of four miles from ground zero, at the edge of the dark gray area in Fig. 6 (top). The integrated thermal energy delivered at that distance at three seconds will be 18 calories per square centimeter.

By this time after the burst, the pulse of neutrons and secondary gammas from the initial nuclear radiation will be over. At ground zero the surface-tissue kinetic energy transfer, or kerma, will be about one gray from neutrons and five grays from secondary gammas. A gray is the Standard International unit for energy deposition by radiation in matter, defined to be one joule per kilogram, or 100 rads, or 10 000 ergs per gram. Gamma radiation from the fission products in the fireball will continue to irradiate the ground for about a minute after the burst, after which time the fireball will have risen to such a height that attenuation in air will reduce this radiation to a negligible amount. The total radiation from fission-product gammas at ground zero will be about 20 grays (or 2000 rads). The total initial nuclear radiation at ground zero from the 1 Mt airburst at 7200 ft will thus be about 26 grays. Whole-body exposure to about 4.5 grays is considered to be lethal to 50% of those exposed.

In comparison, the total initial nuclear radiation dose at ground zero from the 12.5-kt Hiroshima burst at an altitude of 1670 ft is estimated to have been about 133 grays, with about 33 of these grays contributed by neutrons.[3] For small-yield weapons, protection against initial nuclear radiation becomes a predominating factor in shelter design.

At five seconds after the burst, the blast wave has reached a radius of about 1.4 miles from ground zero and the overpressure has reduced to 17 psi.

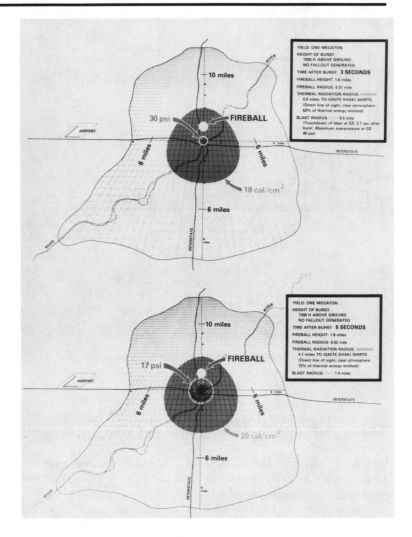

Figure 6. Top: Three seconds after detonation of a 1 Mt bomb on Metropolis.
Bottom: Five seconds after detonation of a 1 Mt bomb on Metropolis.

The blast wave velocity at this stage is about 1.4 times the speed of sound. About 70% of the thermal radiation energy has been emitted (Fig. 7).

At ten seconds after the burst, the blast wave has reached a radius of about 2.8 miles from ground zero and the overpressure has reduced to 9 psi, still strong enough to destroy almost every building (Fig. 8, top). Essentially all of the potentially harmful thermal radiation, about 80%, has been emitted by this time.

At 20 seconds after the burst, the fireball has risen to three miles above the ground, causing a dirty stem of smoke and dust to rise toward it without ever quite reaching it. The fireball is taking on the shape of a doughnut. The blast wave has reached 5.3 miles from ground zero, and now has an overpres-

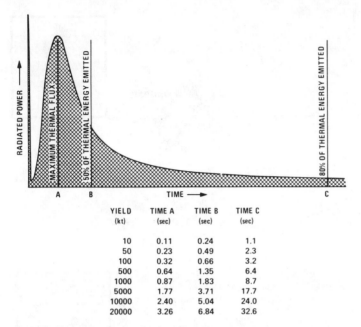

YIELD (kt)	TIME A (sec)	TIME B (sec)	TIME C (sec)
10	0.11	0.24	1.1
50	0.23	0.49	2.3
100	0.32	0.66	3.2
500	0.64	1.35	6.4
1000	0.87	1.83	8.7
5000	1.77	3.71	17.7
10000	2.40	5.04	24.0
20000	3.26	6.84	32.6

Figure 7. Thermal radiation from nuclear weapons.

sure of about 3.5 psi (Fig. 8, bottom). Most brick and concrete buildings with load-bearing walls less than a few stories high remain standing at this overpressure.

At 30 seconds, the blast wave has reached out almost 8 miles from the center, with an overpressure of 1.8 psi (Fig. 9, top). Most buildings remain standing just behind the blast front.

At 55 seconds, the fireball and its stem dominate the picture. The fireball has risen to a height of about six miles, and is beginning to spread out like a pancake (Fig. 9, bottom). The blast wave, with 0.5 psi overpressure, continues to travel outwards, now at a distance of 13 miles and slowed down to the speed of sound. At our position 20 miles away, we will not hear the explosion until about 34 seconds later.

Let us now consider how shelters can protect people from the prompt effects of this detonation. As mentioned before, the overpressure on the ground at ground zero is about 40 psi. Hardly any surface structure, except those constructed like bank vaults, can withstand this overpressure. Most buildings, skyscrapers, brick apartments, and concrete-block structures, will be completely demolished. However, even a homemade shelter buried underground will protect its occupants completely from this overpressure and all the other associated weapon effects.

Figure 10 shows a Soviet design for a shelter constructed out of poles that are obtained by chopping down trees of selected sizes.[4] This shelter is to be constructed in an excavation in the ground and covered with a few feet of earth. Figure 11 shows the Soviet plan adapted for American use. This design

Appendix B   179

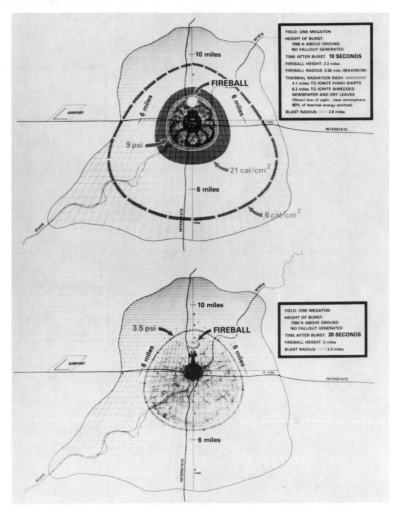

Figure 8. Top: Ten seconds after detonation of 1 Mt bomb on Metropolis.
Bottom: Twenty seconds after detonation of 1 Mt bomb on Metropolis.

has been actually constructed, buried in the ground with five feet of earth cover, and subjected to blast from chemical explosives (Fig. 12). The structure survived (Fig. 13) not just 40 psi overpressure, but 90 psi,[5] and could easily have withstood 100 psi (Fig. 14).[6] If the people in the hypothetical city of Metropolis, or those of Hiroshima, had been in shelters like this one (Fig. 15) at the time of the detonation, there would have been no injuries or fatalities produced by the effects of these weapons. With five feet of earth cover, the dose from initial nuclear radiation inside this shelter directly underneath the Hiroshima burst would be reduced from 133 grays at the surface to less than 0.1 gray.[7]

Even if the people of Hiroshima had been inside the modest shelters they had constructed for protection against conventional bombing, the number of

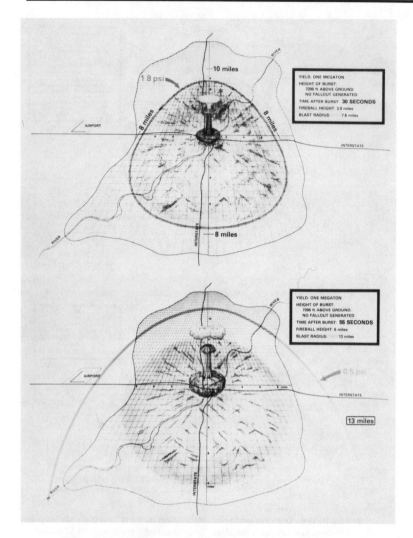

Figure 9. Top: Thirty seconds after detonation of 1 Mt bomb on Metropolis.
Bottom: Fifty-five seconds after detonation of 1 Mt bomb on Metropolis.

fatalities would have been only a small fraction of what it was. These people did not expect a new technology to be used. An air-raid alert throughout Hiroshima at 7 a.m. on 6 August 1945, was called off a half hour later because it appeared to the Japanese that only weather or reconnaissance planes were involved. When the *Enola Gay* arrived with two weather observation planes at 8:15 a.m., no one paid any attention. Many people were working outside, which increased the number of fatalities. Even so, about 193 000 survived, about 55% of the 350 000 people residing in the city at the time.[8] Twenty-six school girls inside a concrete structure only 685 meters from ground zero survived the effects of the weapon.[9]

So far, I have only considered air bursts. There are no simple surface structures that can provide protection from a near-miss ground burst. As may

Appendix B   181

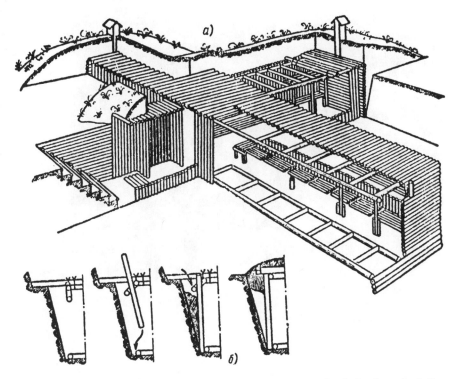

Figure 10. Soviet design for pole shelter.

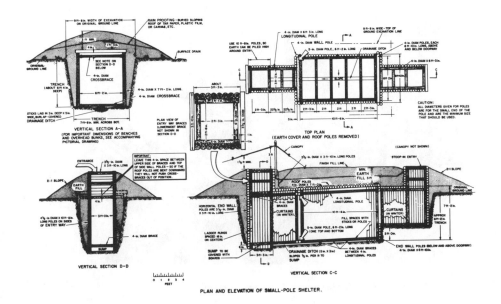

Figure 11. Soviet design for pole shelter adapted for American use.

Figure 12. Chemical explosives over buried pole shelter.

Figure 13. Pole shelter after blast with 40 psi overpressure.

Appendix B 183

Figure 14. Pole shelter after blast with 90 psi overpressure.

Figure 15. Interior of pole shelter.

BLAST OVERPRESSURE ON PLAIN SURFACE FROM SURFACE BURSTS

LOW-MEDIUM OVERPRESSURE RANGE:

$$R = \frac{6 Y^{1/3}}{\sqrt{P} - 0.02(\sqrt{P} - 4)^2}$$

FOR $0.5 \leqslant P \leqslant 500$ psi

HIGH OVERPRESSURE RANGE:

$$R = \frac{3.3 Y^{1/3}}{P^{0.346}} \quad \text{FOR } 500 \leqslant P \leqslant 10{,}000 \text{ psi}$$

FOR R IN MILES
Y IN MEGATONS

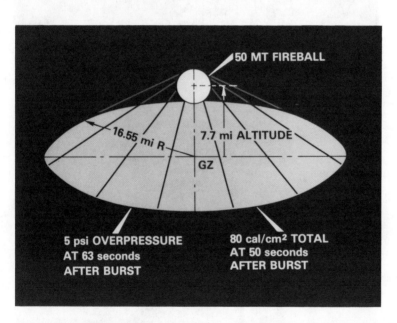

Figure 17. Detonation of 50 Mt weapon in hypothetical situation where population is at highest vulnerability.

be calculated from the equations in Fig. 16, the overpressure at six-tenths of a mile from a 1 Mt surface burst is 100 psi, and the overpressure at about 700 ft is 10 000 psi. These equations were synthesized from data by Harold Brode.[10]

We can use the low-range overpressure equation to compare the areas of lethality to humans between the most vulnerable case and a fairly well protected case. The most vulnerable case is that of completely unprotected per-

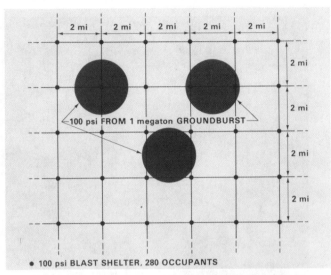

Figure 18. Detonation of nuclear weapons in a hypothetical situation where the population is at lowest vulnerability.

ARTICLES ON FIRES BY A. BROIDO IN *BULLETIN OF THE ATOMIC SCIENTISTS*

1. "MASS FIRES FOLLOWING NUCLEAR ATTACK," vol. XVI, #10, 1960, pp. 409–413

2. "SURVIVING FIRE EFFECTS OF NUCLEAR DETONATIONS," vol. XIX, #3, 1963, pp. 20–23

Figure 19. Articles on fires by A. Broido in *Bulletin of the Atomic Scientists*.

sons who take no evasive action and who are in direct line-of-sight to the fireball. In this situation, thermal radiation at a level of 12 calories per square centimeter from a 1 Mt surface burst will cause third-degree skin burns that will be lethal within a range of 5 miles on a clear day.[11] At this range the overpressure is only 1.8 psi. The area of lethality from the thermal radiation is almost 80 square miles. If people were well protected by being inside 100-psi shelters, this area of lethality would shrink to 1.3 square miles, a reduction by a factor of 60.

It may be instructive to extend this comparison to two hypothetical situations that emphasize extremes in vulnerability of people to the effects of nuclear weapons (see Fig. 17). For the situation of highest vulnerability, suppose

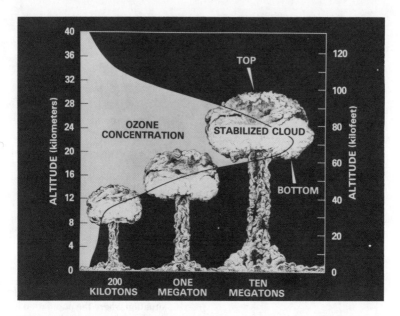

Figure 20. Mushroom clouds generated by ground-burst nuclear weapons.

that all 4.8 billion people of the world are assembled on a bare, flat plain. If we allow 5 square feet per person, which is typical of an outdoor, rock-music festival, all the people of the world can fit inside a circular area with a diameter of 33 miles. A single nuclear weapon of 50 Mt yield would be lethal to this entire population if detonated at an altitude of about 7.7 miles above the center of the circle, and if the atmosphere were clear. The primary cause of lethality would be thermal radiation. The Soviets tested a 58 Mt weapon at an altitude of about 12 000 feet over Novaya Zemlya in 1961.[12]

Now let us consider a situation of the opposite extreme (Fig. 18). For a hypothetical situation of low vulnerability, let us suppose that the U.S. population of the 48 contiguous states, 225 million, is sheltered in equal-size 100-psi blast shelters that are uniformly distributed throughout the land area. There would be about 280 people in each of about 810 000 shelters, spaced two miles apart. In this situation, 10 000 nuclear weapons, accurately delivered and all reliable, could not destroy more than 1.3% of this population, even if the weapons were as large as a megaton each.

I will now discuss the protection that can be provided by shelters against large fires and firestorms. In the *Congressional Record* and elsewhere,[13,14] statements by various members of the group called Physicians for Social Responsibility, have appeared claiming that, even though shelters may protect people from blast, the shelters will, they say, become crematoriums for the occupants when firestorms build up.

There are two responses to this statement. First, it is no great engineering task to design and construct shelters that will protect their occupants from a large fire or firestorm. Even the small-pole shelter described earlier

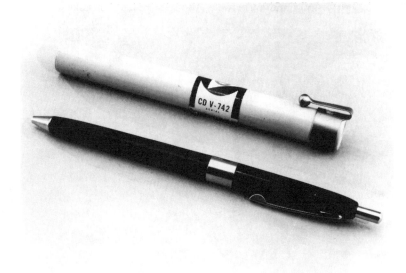

Figure 21. Dosimeter for measuring gamma radiation exposure.

can be made safe for its occupants through the most severe firestorm with just a few precautions.

One of the worst firestorms in the history of mankind took place in Hamburg on 27 July 1943. Within 20 minutes after the incendiary bombing by waves of B-17 bombers, over two-thirds of all the buildings in a five-square-mile area were ablaze. Within a few hours the fire had begun to run out of fuel and die down. Despite the ferocity of this firestorm, official records show that over 85% of the 280 000 people in the firestorm area survived, and nearly all who sought refuge in bunkers, covered trenches, and other nonbasement shelters survived. Many of those 15% who died sought refuge in shelters in the basements of many-storied, heavy-timbered German structures, where they were first asphyxiated and then cremated.[15]

The lessons from the Hamburg evidence are these: (1) people can survive and have survived the worst firestorms, and (2) shelters should not be located where they may be covered by burning structures or rubble unless they have the capability to be sealed off and can provide an adequate supply of air for the occupants for several hours.

The second response to the statement that shelters will become crematoriums is that there are a number of effective civil defense measures that can be taken to greatly reduce the probability of the development of large fires or firestorms after a nuclear attack. This factor was not taken into account as a possibility by the developers of the "nuclear winter" scenario. Rather than describe these measures here, I will refer to two informative articles on the subject by Abraham Broido in the *Bulletin of Atomic Scientists* (Fig. 19).[15,16]

Let us now consider the problem of radioactive fallout from ground bursts. The 1 Mt detonation over Metropolis and the Hiroshima and Nagasa-

188   Civil defense

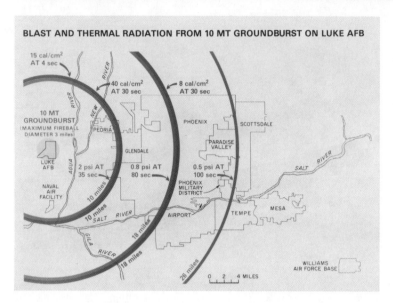

Figure 22. Blast and thermal radiation from 10 Mt ground burst on Luke Air Force Base.

ki bursts were air bursts in which the fireball did not touch the ground. In these cases there is little or no radioactive fallout. After Hiroshima a light rain deposited some radioactive materials on the ground, but it was not enough to cause injury to anyone. The fireball must touch the ground in order to produce dangerous fallout.

When a nuclear weapon explodes near the ground, tons of earth are instantly vaporized and additional tons are pulverized into fine dust which are drawn up into the air with the fireball as it rises (Fig. 20). When the top of the mushroom cloud spreads out and cools, a dust cloud is formed that is carried for miles by the wind. Dust drifts down from this cloud as fallout. These dust particles are highly radioactive mostly from materials from the bomb that have become stuck to the material drawn out of the crater. Gamma radiation from these radioactive materials can penetrate ordinary walls and can be lethal.

Civil defense radiological instruments, such as the dosimeter shown in Fig. 21, measure gamma radiation exposure in units of roentgens. For gamma radiation from fallout, the surface-tissue dose in rads is almost numerically equivalent to the exposure in roentgens. A whole-body exposure to about 50 roentgens within a period of a week may cause some people to feel nauseated, but most adults might not feel anything until they are exposed to about 100 roentgens. According to publication No. 42 of the National Council for Radiation Protection and Measurements, no medical care will be required for normal adults unless a whole-body exposure of 150 roentgens or more is received within a week.[17] Fifty percent of those exposed to 450 roentgens in a week may die, and nearly all of those exposed to 600 roentgens in a week would probably die.

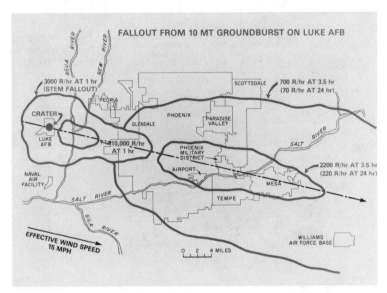

Figure 23. Fallout from 10 Mt ground burst on Luke Air Force Base.

Figure 22 shows some of the blast and thermal radiation effects on the Phoenix, Arizona area if a 10 Mt weapon were ground burst on Luke Air Force Base, about 19 miles from the center of the city. There would be many broken windows, and the destruction of some light wood-frame and sheet-metal structures in the areas of the city closest to the burst. However, the greatest danger to the citizens of Phoenix would arise from the fallout from this weapon if the wind conditions produced the pattern of deposition shown in Fig. 23. The total gamma radiation exposure in a week to an unsheltered person at the outer contour on the slide would be about 6600 roentgens, certainly highly lethal to all mammals out in the open. Again, the same shelters that protect against blast, fire, and initial nuclear radiation will provide protection against the gamma radiation from fallout.

The levels of radiation from fallout will fade away very quickly. During the first few months after deposition, the composite radiation from the dozens of isotopes in the fallout decays with time $T^{-1.2}$. [18,19] This behavior is approximated by what is called the 7:10 rule (Fig. 24). In the specific situation depicted at Phoenix, the radiation intensity of 700 R/h measured at 3.5 hours after the explosion will fall to about 70 R/h in 24 hours, and to 7 R/h in a week after the explosion.

People can be protected from fallout radiation and from the gamma component of initial nuclear radiation by having anything massive, a brick wall, earth, or concrete floors, between them and the source of the radiation. Many buildings will provide such protection without modification. In Fig. 25 several different kinds of buildings and shelters are indicated schematically in order to compare their effectiveness in shielding against fallout radiation. The effectiveness of shielding at the location of a dot shown in a building is indicated by a number called the fallout protection factor, or FPF. When this number is divided into the number estimated for the outside gamma radiation exposure,

190   Civil defense

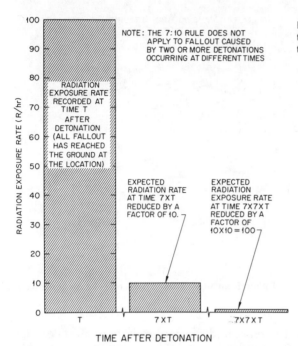

Figure 24. Calculation of decay time of isotope fallout according to the 7:10 rule.

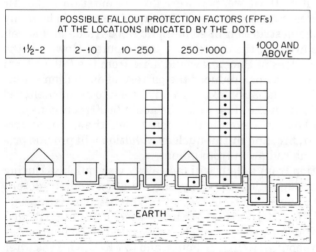

Figure 25. Effectiveness of different kinds of buildings and shelters in shielding against fallout.

the resultant number gives an indication of the radiation exposure to people inside the structure at the location of the dot.

An ordinary wood-frame house built on a concrete slab, indicated at the left of Fig. 25, might provide a maximum FPF of 2. People inside this structure could expect a one-week exposure of 3300 roentgens at the location in Phoenix where the one-week external exposure was estimated to be about 6600 roentgens. These people would not survive. An FPF of about 44 or better

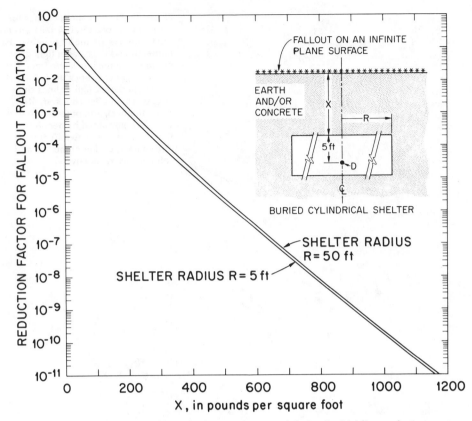

Figure 26. Effectiveness of underground shelter in shielding against gamma radiation.

is required at this location to lower the radiation exposure in one week to a level where no medical care is required, namely, 150 roentgens.

As shown by Fig. 26, it is possible to obtain very good protection from surface gamma radiation by constructing a shelter underground.[20] A special shelter, or the small-pole shelter shown earlier, buried under seven feet of earth, and with a properly designed entrance, will provide so much shielding from even the worst fallout conditions that the radiation received by the occupants from the fallout is less than they would receive from normal background radiation at the surface of the Earth.

I have reviewed the three primary effects of nuclear weapons: blast, thermal radiation, and radioactive fallout from nuclear weapons. I have shown that shelters can provide effective defense against these effects. At the cost of roughly $1000 per space, we could provide concrete blast shelters for half of the population of the United States at a cost of about $120 billion,[21] less than half of our defense budget for one year, which is $266 billion for 1984, and less than half of our health and human services budget for one year, which is $288 billion for 1984. Most of these blast shelters would be located inside urban

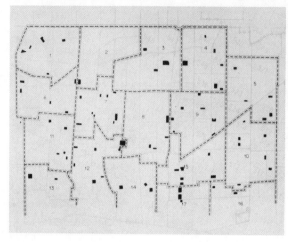

Figure 27. Plan for shelter zones for the inner city of Detroit. 1970 Census tract numbers are the large light gray numerals and the smaller light gray numbers are the block numbers. The heavier light gray lines show Census tract boundaries, lighter lines show streets. The shelter zones, shelter zone numbers, and potential shelter sites, as determined by Bechtel Corporation, are shown by the dotted lines, darker numerals, and solid blocks, respectively.

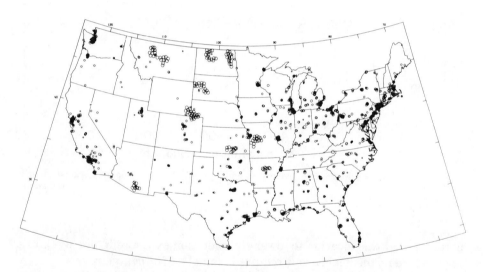

Figure 28. Hypothetical nuclear attack for crisis relocation planning. Circles show areas covered with 2 psi or greater overpressure from blast. Number of delivered weapons: 1444. Total yield delivered: 6559 Mt.

areas. A study of the inner city of Detroit by Bechtel Corporation in 1968 showed that adequate locations could be found so that the average distance to shelter was one-half mile, and the locations would not be covered with rubble after a nuclear attack (see Fig. 27).[22,23]

Of course a 100-plus-billion-dollar, blast-shelter program could not be started instantly. We would have to spend these billions of dollars over a period of several years. The most urgent task of any civil defense program for this country is to educate people on defense against nuclear weapons. Some additional money would be required to make a total civil defense system, which would include fallout shelters for the rest of the population, education

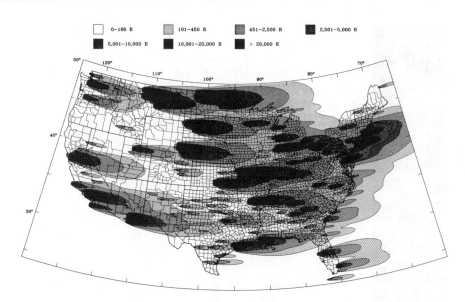

Figure 29. Fourteen-day radiation exposure doses. Effective wind: 20 mph from the west.

of the public, warning systems, stocks of food, water, medical supplies, radiological instruments, protection of communications, and training of special cadres such as shelter managers and radiological monitors.

So far, I have only considered the effects of single nuclear detonations and defense against them. Now let us consider the larger picture where thousands of nuclear weapons are detonated inside the United States. Figure 28 shows circles of 2 psi overpressure from an attack of 10 000 Mt of which two-thirds of the weapons are reliable so 6600 Mt actually detonate throughout the United States.[24] Seventy to eighty-five percent of the U.S. industrial capability would be destroyed by this attack,[24] leaving a manufacturing capability approximately equivalent to that which this country had at the beginning of World War II.[25] Without any civil defense, essentially where we stand now, about 60%–80% of our population would perish, but about 40–80 million persons would survive without trying.[21,24] With a civil defense program, at least 80%–90% of the population would survive, resulting in at least 100 million more survivors than if there were no civil defense.[21] More important, such a civil defense program might deter the would-be attacker so there would not be any attack at all.

If the country were attacked, despite the deterrent of a good civil defense program, much of the nation would be covered with radioactive fallout, and large areas would still be dangerous after a year. In this case the wind is assumed to be from the west at all locations (Fig. 29). This wind pattern results in almost the worst possible fallout for the United States because the country extends east to west more than north to south. I believe from my studies[26] that, despite the claims of the nuclear winter proponents, crops

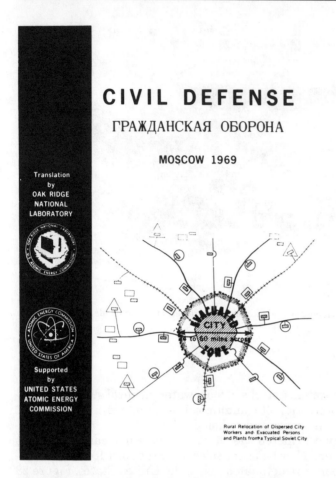

Figure 30. Frontispiece from Soviet volume on civil defense.

could be planted in over 80% of the agricultural areas within a few months to a year after the attack.

According to President Reagan's speech of 23 March 1983, it may be possible to build an active defense that would prevent most nuclear weapons from reaching the United States, a defense that may deter such an attack in the first place. It would be necessary to back up such a system with a strong civil defense.

Now let us examine what the Soviets are doing about civil defense (Fig. 30). Their accelerated efforts on civil defense began about 15 years ago. In recent years it is estimated that they have been spending thirty to forty times more per year on civil defense than the United States, spending possibly as much as six billion dollars on civil defense in 1982.[27] This cost does not include the money they put into missile defense against our bombers and ballistic missile defense. Their civil defense program includes extensive instruction

FOR EXTERNAL PUBLICATION

Radio Moscow in Mandarin to China, Nov. 3, 1978.

"However, the fact is that China's digging deep tunnels can never protect the Chinese masses from nuclear bombing or even protect them from conventional heavy bombs."

* * * * * * * * *

Radio Moscow World Service in English, Nov. 16, 1978

"The U.S. Administration is going to launch a 5-year program of civil defense. - - - The only real safety for the Americans is strengthening friendship with the Soviet Union, not bomb shelters."

* * * * * * * * *

Figure 31. Soviet messages on civil defense composed for consumption outside the Soviet Union.

FOR INTERNAL PUBLICATION

Moscow Voyennyye Znaniya in Russian No. 5, May 1978, p. 33.

"It is appropriate to say that we still meet people who have an incorrect idea about defense possibilities. The significant increase in the devastating force of nuclear weapons compared with conventional means of attack makes some people feel that death is inevitable for all who are in the strike area. However, there is not and can never be a weapon from which there is no defense. With knowledge and the skillful use of contemporary procedures, each person can not only preserve his own life but can also actively work at his enterprise or institution. The only person who suffers is the one who neglects his civil defense studies."

* * * * * * * * *

Figure 32. Soviet message on civil defense composed for consumption inside the Soviet Union.

and training programs for everyone; an evacuation plan to move much of the population of major cities into outlying fallout shelters where food is stored[28]; blast shelters for over half of the working population, perhaps as many as 45 million spaces; and numerous "mirror factories," that is, duplicate factories, some hidden under residential structures in villages.[29] Some 150 000 people are engaged full time as civil defense cadre in a national program that extends from the national level under the Defense Ministry down to cities, rural areas, and industries. An additional 16 million cadre are available under mobilization.[30] It has been estimated that if the Soviets implemented their civil defense evacuation plan, an operation that would take seven to ten days (not overnight, as implied in the movie *The Day After*), an all-out counterattack by the entire U.S. strategic arsenal of nuclear weapons would result in less than ten million Soviet fatalities.[31]

An examination of Soviet radio broadcasts and publications reveals that they have two messages with opposite meanings concerning civil defense. For external consumption the message is that civil defense is useless. For internal consumption, the message is that civil defense is effective and necessary.

Figure 31 shows two examples of messages for external consumption, the first for the Chinese, and the second for the Americans.

Several studies show that there has been a massive expansion of the Soviet strategic disinformation effort in the 1980s.[31–34] One dimension of Soviet strategic deception is to "convince the West that...preparation for war is a meaningless pursuit." [37]

Now let us compare this with Fig. 32 which shows a typical Soviet message for internal consumption, which is completely opposite in meaning to those presented for external consumption.

References

1. *The Effects of Nuclear Weapons*, 3rd ed., edited by Samuel Glasstone and Philip J. Dolan (U.S. Department of Defense and Energy, U.S. GPO, Washington, DC, 1977).
2. Reference 1, p. 36.
3. Tissue kerma vs distance relationship for initial nuclear radiation, George D. Kerr, Joseph V. Pace, III, and William H. Scott, Jr., Oak Ridge National Laboratory, Report No. ORNL/TM-8727, 1983.
4. P. T. Egorov, I. A. Shlyakhov, and N. I. Alabin, Oak Ridge National Laboratory Report No. ORNL-TR-2793, 1973, p. 89 [available from National Technical Information Service (NTIS)].
5. Cresson H. Kearny, Conrad V. Chester, and Edwin N. York, Report No. ORNL 5541, 1980 (available from NTIS).
6. Conrad V. Chester (private communication).
7. C. Eisenhauer and L. V. Spencer, *Approximate Procedures for Computing Protection from Initial Radiation, I and II* (National Bureau of Standards, Washington, DC, in press).
8. *Hiroshima and Nagasaki, The Committee For the Compilation of Materials on Damage Caused by the Atomic Bombs in Hiroshima and Nagasaki*, translated by Eisei Ishikawa and David L. Swain (Basic Books, New York, 1981), pp. 353 and 364.
9. Francis X. Lynch, Science **142**, 665 (1963).

10. Harold L. Brode, Annu. Rev. Nucl. Sci. **18**, 153 (1968).
11. Reference 1, pp. 562 and 565.
12. Reference 1 (1964 edition), p. 681a.
13. Testimony of H. Jack Geiger, M.D., Logan Professor of Community Medicine, City College of New York City, U.S. Congress Senate Committee on Labor and Human Resources, Subcommittee on Health and Science Research, Hearing Short and Longterm Health Effects on the Surviving Population of a Nuclear War, 96th Congress, Second Edition, 19 June 1980, p. 27.
14. Helen Caldicott, Family Weekly **XXX**, 5 (1982).
15. A. Broido, Bull. At. Sci. **19**, 20 (1963).
16. A. Broido, Bull. At. Sci. **16**, 409 (1960).
17. *Radiological Factors Affecting Decision-Making in a Nuclear Attack* (National Council on Radiation Protection and Measurements, Washington, DC, 1974), NCRP Report No. 42.
18. K. Way and E. P. Wigner, Phys. Rev. **73**, 1318 (1948).
19. Edward C. Freiling, Glenn R. Crocker, and Charles E. Adams, in *Radioactive Fallout from Nuclear Weapons Tests*. Proceedings of the 2nd Conference, Symposium Series 5 (AEC Division of Technical Information, Germantown, MD, 1965), pp. 1–43.
20. Carsten M. Haaland, J. Civil Defense **16**, 10 (1983).
21. Roger J. Sullivan, Winder M. Heller, and E. C. Aldridge, Jr., System Planning Corporation Report No. 342, 1978.
22. *Protective Blast Shelter System Analysis—Detroit, Michigan* (Bechtel Corporation, 1968).
23. Carsten M. Haaland and Michael T. Heath, Oak Ridge National Laboratory Report No. ORNL-4818, 1972.
24. Carsten M. Haaland, Conrad V. Chester, and Eugene P. Wigner, Oak Ridge National Laboratory Report No. ORNL-5041, 1976.
25. *The Statistical History of the United States From Colonial Times to the Present* (Basic Books, New York, 1976), p. 666..
26. Carsten M. Haaland, Am. J. Agri. Econ. **59**, 358 (1977).
27. Leon Gouré, *Soviet Civil Defense Concepts, Programs and Measures for the Protection of Industry in Nuclear War Conditions* (Advanced International Studies Institute, 1981).
28. Carsten M. Haaland and Eugene P. Wigner, Natl. Rev. **28**, 1005 (1976).
29. Robert Edward Kirshensteyn, J. Civil Defense 16 (1983).
30. *The Military Balance 1983–1984* (The International Institute for Strategic Studies, London), p. 15.
31. Eugene P. Wigner, Survive **3** (1970).
32. Vladimir Bukovsky, Commentary, 25 (1982).
33. Leon Gouré and Michael J. Deane, Strategic Rev. **11**, 79 (1983).
34. Leon Gouré and Michael J. Deane, Strategic Rev. **11**, 81 (1983).
35. J. A. Emerson Vermaat and Hans Bax, Strategic Rev. **11**, 64 (1983).
36. Joseph D. Douglass, Jr., Orbis, **27**, 667 (1983).
37. *Director's Report* (Physicians for Social Responsibility, Cambridge, MA, 1983), p. 6.

Appendix C: The U.S. needs civil defense*

Roger J. Sullivan

President Reagan has repeatedly called for a steeply upgraded civil defense program. Congress should support this.

The Soviet Union and the United States currently each possess several thousand strategic offensive nuclear weapons, deployed on intercontinental ballistic missiles (ICBM's), submarine-launched ballistic missiles (SLBM's), and bombers. In principle, three broad types of strategic defensive systems can counter these: antiballistic missiles (ABM's—to shoot down incoming ballistic missiles), air defense (to shoot down incoming bombers), and civil defense (to protect the country's assets, especially population, from the effects of nuclear weapons that penetrate and explode). Soviet air defense and civil defense are each many times as extensive as their U.S. counterparts. (ABM's are essentially prohibited on each side by means of the 1972 ABM Treaty.)

For many years U.S. policy has been based on "mutual assured destruction"—the theory that neither side will attack the other because, if it did, the attacked nation would retaliate and destroy the attacker. Such a principle is highly dubious because (1) it is based entirely on our perceptions of Soviet intentions, not on their capabilities, and it may prove wrong in a crisis; and (2) the Soviets have never subscribed to it. Furthermore, as accuracy improves and the use of multiple independently targeted reentry vehicles (MIRV's) increases, either side may in the midst of some future crisis conclude that, by attacking, it can destroy far more missiles than it need expend, thus possibly making an attack seem worth the cost. Thus the "crisis stability" of the current situation is not encouraging.

The situation would be much more stable if it were based on a principle of "mutual assured survival." Each side would possess relatively low levels of strategic offensive systems, limited by means of arms control negotiations and relatively high levels of strategic defensive systems to reduce the effectiveness of an attack. Stability would be greatly increased because (1) it would be based on verifiable Soviet capabilities, not their intentions or doctrine; and (2) any attacker would have to expend more weapons than he could destroy, thus removing the incentive for an attack. Such a policy requires both offensive arms control and strategic defense—two concepts that are complementary, not contradictory.

*This paper was presented at a *Forum* Symposium on Civil Defense at the March 1984 Meeting of The American Physical Society in Detroit, MI by Roger J. Sullivan, System Planning Corp., 1500 Wilson Blvd., Arlington, VA 22209. This paper is also adapted from an article in the *Journal of Civil Defense* of August 1982.

The most extensive plausible nuclear attack against the U.S. is one that targets our military facilities and industry. However, population *per se* is not considered a target by the Soviets: thus an evacuated population would most probably not be targeted. A map of probable "risk areas" reveals that blast and fire would very likely cover only a few percent of the 48 contiguous states. Because industry is generally located in cities, this area includes about 70% of the population. Such an attack would also produce extensive radioactive fallout over much of the nation. It would blow generally from west to east and decay substantially during the days following the attack. Although an attack could conceivably be launched "out of the blue," most analysts believe it far more likely that it would arise from an escalating crisis over a period of several days or weeks.

Protection against an out of the blue attack would require a nationwide system of blast shelters and would cost over $60 billion. In the current cost-cutting environment such a commitment by the government seems highly unlikely. However, protection against the more likely attack-from-crisis, based on evacuation, would be far less costly: about $2 to $4 billion over five years or about $2 to $4 per American per year (we currently spend about 50 cents per American per year on civil defense).

Four important questions must be addressed regarding such a program.

(1) *Would a crisis relocation program (CRP) work if people cooperated?* Under a CRP, the officials of each state and county, in coordination with the Federal government, would establish detailed plans and preparations for moving people out of primarily urban high-risk counties and into rural low-risk host counties. To protect themselves from radioactive fallout, people would have to establish "expedient" fallout shelters, following instructions provided by the government. This could be done most easily by going into basements of existing buildings and piling earth around the outside, and in some cases on the first floor, to attenuate the radiation. A CRP should include detailed county-by-county preparations for traffic regulation, building allocation, and stockpiling of essential supplies, including water, sanitation kits, medical supplies, and some food. A great many issues have been analyzed concerning evacuation, including key workers to maintain vital functions and prevent looting, traffic control, evacuation of people without cars, fuel supplies, housing and food in the "host" areas, and so forth. The overall conclusion is that, if the people cooperate, it can be accomplished successfully in one to three days. Analyses show that a large-scale attack would kill roughly 80% of the American people if there were no preparation, but about 20 % if the CRP had been successfully implemented in advance.

(2) *Would people cooperate?* Since World War II there have been over 200 evacuations within the United States as a result of actual or impending natural disasters, such as earthquake, hurricane, or flood. These evacuations have been routinely successful. State and local civil defense officials are capable individuals who know how to direct an evacuation without its resulting in injuries or inordinate chaos. Experience shows that the better the advance preparations, the more smoothly the evacuation proceeds.

In the absence of disaster, many people are apathetic or even hostile to the idea of evacuation; but when a real disaster seems imminent, people put aside their predisaster attitudes and cooperate with officials and with each other to a surprisingly high degree. To encourage such cooperation it is particularly important that officials provide as much information to the people as possible, before and after the disaster.

Whereas a natural disaster is an island of disaster in a sea of normalcy, an impending nuclear attack would imply potential disaster areas all across the country, corresponding to all military facilities and sizable cities. Extensive peacetime preparations, coordinated among Federal, state, and local officials, would be a prerequisite to an orderly nationwide evacuation. Nevertheless, to any particular individual one type of evacuation would appear about like the other. Thus it is reasonable to expect people generally to cooperate with authorities during a nuclear crisis evacuation. Furthermore, experience (e.g., Cuba 1962, Three Mile Island) has shown that, during a perceived crisis, a substantial fraction of the people will *spontaneously* evacuate. If for no other reason, nationwide CRP is necessary to channel such spontaneous evacuation and help people to relocate to relatively safe areas.

In a nuclear crisis evacuation, people would have to follow instructions from authorities to establish expedient fallout shelters in the host areas. Over the years many tests have been conducted with untrained individuals, and have demonstrated that, given information and incentive, people can and will construct such shelters and live in them for several days.

Polls have shown that over 75% of the American people want good civil defense and are willing to pay the cost of a CRP. Most people do not become actively interested in civil defense until a crisis occurs; however, at that point they besiege the government for information and instructions and expect government to be ready to provide leadership.

(3) *Would postwar survival and recovery be possible*? A number of detailed studies have concluded that, if people are sheltered until the radioactive fallout decays, and if proper preparations are made for continuity of government and management of surviving resources, then long-term survival and recovery are indeed possible. An excellent book on how individuals can protect themselves is *Nuclear War Survival Skills* (Caroline House, Naperville, 1980) by Cresson H. Kearny. It explains how to evacuate, construct shelter, obtain safe water, food, light, and sanitary facilities, and how to survive without doctors. I have personally conducted research on survival during the first year after a nuclear attack, considering the availability of such essentials as fuel, transportation and communication facilities, food, water, housing, clothing, sanitation, the threat of disease and long-term radiation, and potential ecological disruptions following an evacuation and a nuclear attack. Life would be considerably more difficult than it is today, and many cities would be in ruins. I concluded, nevertheless, that if governments make sufficient preparations in peacetime, the people can survive in the postattack environment and begin to rebuild the nation.

Recently, a group of scientists has published two papers [*Science* **222**, 1283 and 1293 (1983)] containing the conclusion that the Earth will experience a "nuclear winter" in the event of nuclear war. Related articles have emphasized a "threshold" of 100 megatons for a nuclear winter to occur. This issue is important and deserves further study. In the meantime, however, it is important to remember that (1) some experts disagree with the assumptions and conclusions in the *Science* articles, (2) the 100 megaton "threshold" applies to pure counter-city attacks, which are of essentially zero likelihood, (3) a 3000 megaton counterforce attack (one of the most likely cases) does not produce a nuclear winter, and (4) the second paper (biology) is based on the very worst case of the first paper (climatology), not the "baseline" case, as it should be. In my opinion the potential environmental effects of nuclear war are sufficiently uncertain that they do not diminish the value of civil defense. Our approach should be, not to abandon civil defense, but rather, to adopt preparations for surviving the types of environmental effects that may occur.

(4) *Would a U.S. CRP increase the chance of nuclear war?* In 1978 I interviewed about thirty authorities on crisis management and nuclear strategy—including liberals, conservatives, and "middle-of-the-roaders"—on the question of whether an effective U.S. CRP would be likely to precipitate a serious crisis or make nuclear war more likely. I concluded the following. If a nuclear crisis occurred, it would result from many complex and unpredictable causes. The presence or absence of U.S. or Soviet civil defense would have a relatively minor effect on the central events of the crisis itself and would probably not materially contribute to the chance of escalation to nuclear war. The U.S. should probably not evacuate in the absence of Soviet evacuation. However, if the Soviets begin to evacuate, then we should definitely do so as well, to protect our people if war follows. Such a U.S. responsive evacuation would not be likely to escalate the crisis further, and could well contribute to deescalating it.

Civil defense may be likened to a seat belt in a car. From time to time I achieve a particular objective by driving my car from one place to another. I drive as carefully as I can, and try my best to avoid accidents; however, I also wear a seat belt to minimize the damage to myself should an accident occur. Similarly our country should conduct its national policy so as to achieve our objectives while trying our best to avoid nuclear war. Yet we need the "seat belt" of civil defense to minimize damage to our people should nuclear war nevertheless occur. I believe that civil defense would not increase the chance of war any more than seat belts increase the chance of automobile accidents.

Since World War II the U.S. government has performed over 25 broad studies of civil defense and many hundreds of studies of its various components. The overwhelming conclusion is that it can work and we need it. In my view, the purpose of civil defense is not to make nuclear war more "thinkable" (thousands of people think about it every day) or "winnable" (this is admittedly a dubious concept). It is simply to provide as much protection as possible in case nuclear war occurs. The weapons are there. No physical barrier prevents them from being used. People and nations are unpredictable. A serious superpower crisis can occur anytime. Several days of advance indication of attack might well be available, especially if the Soviets began to evacuate. Heavy spontaneous evacuation would occur in the U.S. The public would cry out for leadership by government. Proper government preparation would provide the American people with a greatly increased chance of survival and recovery should deterrence ultimately fail and nuclear war occur. We have an obligation to ourselves and our descendants to protect ourselves as much as possible against this terrible disaster. We need civil defense.

Appendix D: Under the mushroom cloud*

Barry Casper

Testifying before the Senate Foreign Relations Committee on civil defense in 1982, the former Commander-in-Chief of all U.S. Forces in the Pacific, Admiral Noel Gayler tried to inject a note of reality based on his experience: "I have no confidence in the imaginary situations and chess games that a certain school of analysts dreams up. Real war is not like these complicated tit-for-tat imaginings. There is little knowledge of what is going on, and less communication. There is blood and terror and agony, and these theorists propose to deal with a war a thousand times more terrible than any we have ever seen in some bloodless, analytic fashion. I say that's nonsense."

What Admiral Gayler was referring to is nuclear war crisis relocation. With authoritative support from hired experts, the Federal Emergency Management Agency (FEMA) has been trying to sell the American people on a plan to evacuate the nation's cities and other "high risk areas" in the event the President decides that nuclear war is imminent. Voluminous supportive studies purport to show that with crisis relocation 80% of the population would survive an all-out nuclear exchange and with the construction of blast shelters, the survival rate could reach 90%. The images this program advertises are seductive: "Load your family in your car and drive away from nuclear war,...Dig a hole and you can protect your loved ones from nuclear blasts."

The promoters appeal to reason by stating that despite our best efforts to prevent nuclear war, a large-scale Soviet nuclear attack against the United States is entirely possible; it is only prudent to prepare for that event and save as many lives as possible. They also appeal to fear by stating that if the Soviets have this capability and we do not, they can evacuate their cities and subject us to nuclear blackmail.

As an example of how this program is being sold, consider the preamble of the crisis relocation plan distributed in one western Minnesota city "Today, all Americans face the most serious crisis this country has known. All of our government's attempts to negotiate a peaceful settlement of the differences between the United States and our potential enemies have apparently failed. The specter of a nuclear war and an attack on this country become more and more possible. A nuclear strike affecting any one of the risk areas in Minnesota could be imminent. At this time our diplomats are seeking every

---

*This paper was presented at a *Forum* Symposium on Civil Defense at the March 1984 Meeting of The American Physical Society in Detroit, MI, by Barry Casper, Physics Department, Carleton College, Northfield, MN 55057.

means of avoiding such a devastating holocaust. At any time the President may declare a state of emergency. If and when this happens everyone's cooperation is required to save the lives of as many Americans as possible. *Remember, your life, those of your dependents, and the survival of our country are at stake."*

It has not worked. In risk cities and host towns around the nation, citizens who have examined the crisis relocation plan for their community have decided not to participate. In the past three years, more than a hundred jurisdictions, from New York City to San Francisco, have rejected the plan.

The story of how crisis relocation planning (CRP) has run into trouble is a fascinating case study of experts and ordinary citizens examining a problem from different perspectives. I have some insight into what went wrong with CRP from first-hand experience in my home town of Northfield, Minnesota. In mid-1982, a public meeting there alerted the community to the fact that Northfield was a host area, assigned to shelter evacuees from the Twin Cities risk area, forty miles to the north. The City Council appointed a Crisis Relocation Task Force, headed by the Northfield police chief, which included a housewife, a school bus driver, a psychologist, our local civil defense director, and myself, a physicist. We spent many weeks studying the civil defense literature in preparation for a series of over twenty open meetings with local civic groups. The issues we found ourselves dealing with had some technical elements, but common sense judgments about how people act and political considerations were important factors as well. In the process, I developed a slide show that illuminates some of these issues. It is called *Under the Mushroom Cloud*.

FEMA's experts have contributed to the CRP debate through their research on the consequences of nuclear war and the mitigating effects of crisis relocation. But their generally sanguine predictions about the prospects for surviving a nuclear attack have not been decisive in convincing people to support CRP.

There are two major reasons why the FEMA experts' analyses have not proved to be decisive in many communities. The first reason has to do with the credibility of the plan; in this era, many people have learned to approach technical expert opinion and analyses with a healthy degree of skepticism and, if their common sense conflicts with the experts' conclusions, to trust their own judgment. With civil defense preparations, you are talking about things that are going to happen right there in the community. The people of that community have first-hand experience to draw on in deciding whether what is being proposed is sensible or not.

In Northfield, you could feel the skepticism grow at the first public meeting, when the chief state CRP planner described where Twin Cities evacuees would be housed—complete with 23 pages of computer printout detailing the shelter capacity of every public building and business in town. For instance, the proprietor of Bob's Shoes was absolutely incredulous when he found that 170 people were supposed to fit in his basement, which measured only about 40 feet on a side. And the police chief, a very sensible man, highly respected in the community, was damned sure that evacuation of Minneapolis under the threat of nuclear war would in no way resemble the plan described at the meeting.

After the meeting the Mayor appointed our task force. We read the civil defense studies and met with state and regional civil defense officials. In trying to comprehend the death and destruction that nuclear war would bring, I think it is fair to say that we reacted to the FEMA literature in much the same way as Admiral Gayler. Our intuition was that nuclear war and its aftermath will be much more complicated, much less predictable, much more chaotic, and much more terrible than the sterile paper studies and computer models had grasped.

And the more we understood and thought about the assumptions implicit in crisis relocation, the more implausible it seemed. Three of the assumptions are considered in the slide show: the three days that are supposed to be available between the time the President sounds the alarm and the time the bombs arrive, the orderly evacuation, and the expectation that those who reach the rural host areas will survive.

We were buttressed in our skepticism by a number of authoritative experts like Admiral Gayler, who had weighed in the debate on the other side. A congressional study by Arthur Katz, later published as the book *Life After Nuclear War*, was persuasive in its description of the enormous economic and social disruption that would be caused by crisis relocation itself, even if there were no war, and in its analysis of the breakdown of the fabric of the American economy that nuclear war would cause. A variety of critiques by Physicians for Social Responsibility (PSR) and its local doctors made a compelling case for the inability of our health care system to deal with the overwhelming consequences of a nuclear attack. Dr. Herbert Abrams's article in the *New England Journal of Medicine* on the likelihood of uncontrollable epidemic diseases, and other criticisms of CRP, since assembled and published in the thoughtful PSR anthology *The Counterfeit Ark: Crisis Relocation for Nuclear War*, suggested many damaging and possibly fatal flaws in the FEMA analysis. And the Swedish Academy of Sciences special edition of its journal *Ambio* on nuclear war raised the specter of a postattack period of darkness and cold, subsequently reinforced by the "nuclear winter" studies of Carl Sagan and his colleagues. The FEMA studies seemed superficial and glib by comparison.

For our task force of nonexperts all this added up to a credibility gap that assurances from state and local civil defense officials were unable to patch. In fact, it was our perception that some of those officials shared our doubts, though they were unable to state this publicly.

Beyond the somewhat technical questions of whether or not CRP will "work," there is a second, equally important, political reason why planning for nuclear war evacuation has encountered nationwide resistance. Discussion of crisis relocation is inextricably intertwined with the great debate now taking place over nuclear weapons policies. In at least two ways, CRP can be regarded as an effort to prepare for nuclear war. First, of course, is its humanitarian function of saving lives if we are attacked. But, in the minds of some military planners, CRP also has an important strategic function. They argue that despite our best efforts, nuclear war may happen and if it does happen, we want the United States to "win"; "prevail" is sometimes the way they put it. To that end, we need a nuclear-war-fighting capability, including the capacity to strike the Soviets first if we become convinced we are about to be attacked. That means we need counterforce weapons like the MX, Trident

II, and Pershing II to destroy Soviet missiles and bombers before they are launched; effective programs of antisubmarine warfare to take out their strategic submarines; an active defense against the missiles and bombers that survive; and crisis relocation to evacuate our cities. CRP is an essential element in a nuclear-war-fighting capability.

In the current political debate over nuclear weapons policies, CRP has become a central symbol of what is widely perceived as an intensified effort to "prepare for" nuclear war. Because it is the part of the preparation effort that directly touches people where they live, it has become a lightning rod for expressions of public concern, as well as an opportunity for effective political action.

In the CRP discussion in Northfield, for instance, a major issue that emerged was whether local energies and resources would best be used in preparing for nuclear war through crisis relocation planning or for preventing nuclear war by challenging our present nuclear weapons policies. It was decided that by rejecting the plan, our community in rural Minnesota could send a political signal to Washington that we want no part in an ultimately futile, potentially expensive, and possible dangerous effort to prepare for nuclear war.

After six months of study and community-wide discussion, in December 1982 the Northfield City Council unanimously rejected participating in CRP and instead authorized a Mayor's Task Force on Nuclear War Education and Prevention. That task force is hard at work right now.

To say that we need to pursue both nuclear weapons arms control *and* nuclear war civil defense is misleading and naive. It ignores both the essential strategic role of CRP in a nuclear-war-fighting capability and the important political role CRP has taken on in a historic public debate.

Appendix E: Nuclear winter and the strategic defense initiative*

Caroline Herzenberg

This communication discusses the capability of large-scale directed energy missile defense systems in space for causing extensive fire damage if redirected to surface targets, and the potential climatic effects of such fires. The study was undertaken because of the importance of the Strategic Defense Initiative in the national security and arms control areas, and concern that recent analyses in the literature of the advantages and disadvantages of ballistic missile defense systems have had too narrow a focus, addressing only the potential of such systems within the context of their intended application as defensive systems. The conclusions are that such large-scale space ballistic missile defense systems employing high-intensity lasers operating at frequencies at which the atmosphere is substantially transparent may have the potential for causing devastating surface fires so massive that severe climatic effects similar to those addressed in nuclear winter calculations may ensue.

Since 1982, there has been extensive discussion in the literature of the global atmospheric effects of nuclear war, and, in particular, of the possibly catastrophic climatic effects of smoke generated in a nuclear war, the "nuclear winter" effect.[1-3] It would appear to be important to examine large-scale non-nuclear weapons systems to assess their potential effects in ameliorating or exacerbating a nuclear winter, or indeed even their potential for independently causing severe climatic effects similar to those of nuclear winter.

Since the advent of the concept of the Strategic Defense Initiative (SDI) in 1983, it has been clear that this ballistic missile defense technology offers the possibility of limiting the effects of nuclear winter even in a large-scale nuclear exchange as a collateral effect of ballistic missile defense. This paper directs attention to the circumstance that alternative utilization of a large-scale defensive weapons systems similar to those envisaged in the context of the SDI or similar Soviet systems might in fact have the deleterious effect of leading to nuclear-winter-like effects independent of the occurrence of a nuclear war.

SDI has been characterized as "a new research program...to study how lasers, particle beams, and homing projectiles could destroy ballistic missiles to protect populations against a massive first strike."[4] SDI is based to a signif-

*This paper is reprinted from *Physics and Society* 15(1), 2 (1986). The author is Caroline Herzenberg, Argonne National Laboratories, 9700 South Cass Ave., Argonne, IL 60439.

icant extent on directed energy missile defense in space, although recently kinetic energy weapons concepts have received attention.[4-6] The possibility of initiating severe climatic effects similar to those of a nuclear winter through use of a large-scale directed-energy missile defense system arises because lasers could be employed in a manner not originally envisaged in SDI, as has been discussed in a recent informal report by Latter and Martinelli:[7] "SDI: Defense or Retaliation?" Specifically, the laser weapons of such a large-scale system, instead of being targeted against offensive missiles and reentry vehicles in space, could be directed against targets on the ground, including cities. Rough calculations indicate that a laser defense system designed to be powerful enough to cope fully with the ballistic missile threat posed by either superpower might also have the potential of initiating massive urban fires and even of destroying the enemy's major cities by fire in a matter of hours. Such mass fires might be expected to generate smoke in amounts comparable to the amounts generated in some major nuclear exchange scenarios. Since it is primarily the effects of the smoke generated by fire in a nuclear war which lead to nuclear winter, it appears that a climatic catastrophe similar to nuclear winter might also result from such a ground-target-directed application of intense laser beam weapons from a large-scale system similar to a ballistic missile defense system.

What is the technical basis for such a judgment? We limit our attention to lasers, and exclude from consideration other weapons proposed for the arsenal of SDI, such as x-ray lasers, kinetic energy weapons, neutral particle beams, and laser-channeled electron beams.[4-6,8] Only space-based or ground-based lasers, operating at wavelengths at which the atmosphere is substantially or appreciably transparent, are relevant. (However, since systems information available in the unclassified literature is rather limited, data on laser weapons systems operating at other wavelengths will also be used for illustrative purposes.)

It is necessary to consider some specific examples to provide quantitative information to support the inferences stated earlier. A laser weapon causes damage by concentrating thermal energy on its target in excess of what the target could withstand without malfunctioning.[8] Various informed sources report that a laser beam applied against a missile must direct of the order of 10 000 J/cm^2 of radiant energy for a time interval of the order of seconds in order to achieve such damage by thermal kill.[4,6,8] In fact, delivery of a burst of laser energy of 10 000 J/cm^2 to burn through a missile skin in one second at a distance of 3 000 km has been described authoritatively as a reasonable level of lethality and an acceptable range in terms of the size of the constellation of platforms to deal with a massive attack.[4]

It should be noted that the radiant energy exposure of 10 000 J/cm^2 required for a reasonable level of lethality against missiles far exceeds the radiant energy exposure needed for ignition of fires. The report by Turco et al. on nuclear winter defines the area of urban fire ignition by the 20 cal/cm^2 (80 J/cm^2) contour,[1] more than two orders of magnitude smaller than the missile lethality value quoted above. Radiant exposures from nuclear weapons thermal radiation as low as 16 J/cm^2 are reported to cause ignition of fires in some common household and outdoor tinder materials.[3,9]

Let us now examine the capability of a large-scale defensive system similar to what has been envisaged in connection with SDI for employing high-intensity lasers to initiate mass fires. Parameters appropriate to an orbiting chemical laser boost-phase intercept defense system using HF chemical lasers, radiating at 2.7 μm in the infrared, have been treated in the most detail in the open literature.[6] (This case would not be suitable for the incendiary attack mode under consideration here. Infrared radiation of this wavelength is attenuated by the atmosphere; however, most of it gets down to 10 km or so. Such lasers can be considered for a boost-phase intercept system.)[6]

Such a system might consist of 160 separate 20-MW HF chemical lasers, with output optics consisting of 10-m mirrors, orbiting at 1 000 km altitude.[6] The minimum divergence angle for the resulting beams would be 0.32 μrad.[6] The spot from such a laser beam at a range of 4 000 km would be 1.3 m in diameter.[6] Twenty megawatts distributed over this spot size would give an average energy flux of 1.5 kW/cm^2. Thus to irradiate a missile at this range at the nominal lethal fluence would require a target dwell time of 6.6 sec. However, to irradiate at the nominal 80 J/cm^2 for ignition of fires would require a dwell time of around 50 msec. At a range of 2 000 km, perhaps more realistic, the dwell time for ignition of fires would be about 13 msec.

Each such laser would be designed to be operable for at least 150 sec, the accessibility time for boost-phase intercept for boosters resembling the U.S. MX missile.[6] Thus, assuming a negligible slewing time of the beam from target to target, each such laser could deliver the nominal fluence for ignition of fires at over 10 000 separate locations during the 150 sec of total use. (These are order-of-magnitude numbers only. For more details, see the Supplement.)

What would be the efficacy of 10 000 separate simultaneous ignition points within a city in creating mass fires, such as conflagrations or in particular a firestorm? Experience during World War II may provide some guidance.[9,10] Conflagrations, as distinct from firestorms, are mass fires having moving firefronts which can be driven by the ambient wind. The fire of a conflagration can spread as long as there is sufficient fuel. Conflagrations can develop from a single ignition, whereas fire storms have been observed only where a large number of fires are burning simultaneously over a relatively large area.[9]

In a firestorm, many fires merge to form a single, convective column of hot gases rising from the burning area and strong, fire-induced, inwardly radially directed winds are associated with the convective column. The conditions under which a firestorm may be expected are not well known. However, based on World War II experience with mass fires in Germany and Japan, the minimum requirements for a firestorm to develop are considered by some authorities to be the following: (1) at least 8 lb of combustibles per square foot of fire area, (2) at least half of the structures in the area on fire simultaneously, (3) a wind of less than 8 mph at the time, and (4) a minimum burning area of about half a square mile.[9] Since urban flammable material burdens average 10 g/cm^2 in city centers,[1] corresponding to about 20 lb/ft^2, under suitable weather conditions there is the potential for creating a firestorm if more than half of the structures in an area of half a square mile or more can be ignited simultaneously.

Staying with actual experience, the Hamburg firestorm was created during World War II when about 700 bombers in two attack phases dropped approximately 2 400 tons of mixed incendiaries and high-explosive bombs on the eastern and southeastern districts of Hamburg.[10] The Overall Report (European War) on the U.S. Strategic Bombing Survey states that two out of three buildings were afire within a 4.5 square mile area, while general fires were started over a total area of about 17 square miles.[10] In less than one hour, the fires in the core area had merged to form a mass fire; the total core of the firestorm embraced almost 4.5 square miles.[10]

If we use these numbers, we can make an order-of-magnitude estimate of the number of initiating points for firestorm formation. Allowing 30 buildings to the 1/8 mile city block, one finds roughly 6 000 equivalent initiating points for the Hamburg firestorm. The minimum criteria for development of a firestorm mentioned earlier appear to require only on the order of 500 initiating points. Thus the capability of creating more than 10 000 simultaneous ignition points for fires appears fully adequate for creating a firestorm or conflagrations in an urban area under appropriate weather conditions. So it appears that a single laser battle station could have the capability of destroying a city by incendiary attack.

For a ballistic missile defense of the type that has been considered in connection with SDI, there would be a constellation of a large number of such battle stations deployed in orbit above the Earth.[6,11] In the example under consideration, there would be 160 separate lasers of the type just examined orbiting in the base case (or possibly even 10 times as many should arsenals increase, or the system be enhanced to cope with fast-burn boosters).[6] Thus substantially all of the major cities of either superpower could be targeted for radiative thermal attack by intense lasers, with the potential for creating mass fires in all of these urban areas within a matter of hours.

It should be noted that the estimates above take into account only the potential of the boost-phase portion of a large-scale ballistic missile defense system for incendiary attack application. Laser systems associated with post-boost, midcourse, and terminal defense might enhance the incendiary capability of a large-scale missile defense system by an order of magnitude or more.

Prior calculations[7] using somewhat different assumptions, and considering other types and configurations of lasers, indicate the possibility of even greater efficacy of a large-scale laser ballistic missile defense system for incendiary attack, and suggest up to a total of 100 million separate ignition points.

Could severe to catastrophic climatic effects like those of nuclear winter follow such a laser attack? Turco *et al.* found nuclear winter effects even in their 100-Mt city attack scenario, an attack which corresponded to burning 100 major cities with no other significant destruction.[1] Since an attack on cities by a large-scale laser weapons system similar in characteristics to those associated with directed energy missile defense systems appears to have the potential of creating massive urban fires in over 100 cities within a matter of hours, there appears to be a serious possibility that a nuclear winter could ensue directly from such a large-scale laser attack, without a nuclear war.

Supplement: Notes on attenuation and beam divergence

When laser radiation is transmitted through the Earth's atmosphere, numerous physical processes can occur which, generally speaking, alter the nature of the beam. In particular, the beam will be attenuated by absorption and scattering processes. Absorption and scattering result not only from constituent gases of the atmosphere, but also from water vapor, water droplets, and other particulates including smoke and dust. The absorption and scattering cross sections are functions of wavelength; and attenuation of electromagnetic radiation in the Earth's atmosphere is strongly wavelength dependent.[12,13]

The total beam attenuation is also of course dependent on the total path length through the atmosphere. For the case of a laser beam originating from an Earth-orbiting satellite and directed at a ground target, the minimum total path length through the atmosphere would occur for the case in which the laser is directed vertically downward (air mass 1, zenith angle 0° as seen from the target).[12,14] For this case, the Earth's atmosphere would be rather opaque at some wavelengths; however, in certain other wavelength ranges, most notably in the visible as well as in several bands in the infrared, the Earth's atmosphere is largely transparent, and the transmission of electromagnetic energy vertically down through the entire atmosphere can in some of the regions, under suitable conditions exceed 75%.[13,14]

Not only can transmission be adequately large for a laser beam directed vertically downward from a satellite battle station; but also, even distant surface targets can be reached with appreciable beam transmission. This is because, for satellites at the altitudes under consideration, most of the beam path is in the vacuum of space, so that targets at even comparable or larger horizontal distances can be irradiated without great increases in the air mass traversed by the beam.

Thus, for example, a satellite battle station (or relay satellite) located 1 000 km above the Earth's surface could direct a laser beam to attack a target at a slant range of 2 000 km; under these conditions (zenith angle approximately 60° as seen from the target location; target located approximately 1 700 km away from directly beneath the satellite), the air mass traversed by the laser beam would be only approximately twice as great as that traversed by a vertical beam; and the transmission of the beam would still exceed 60%.[13,14] Propagation of a laser beam in the Earth's atmosphere will also be affected by atmospheric turbulence, and, for the case of high-power laser beams, by phenomena such as laser blooming as well.[15,16]

Microscale temperature fluctuations, which are due to turbulent mixing, cause the refractive index of the atmosphere to vary by parts per million as a random function of position and time, affecting optical wave propagation in the atmosphere.[15] Under moderate turbulence, the focal plane distribution of the laser beam retains its diffraction limited beam size, but moves randomly under the influence of the large turbulent scale sizes; in the presence of strong turbulence, the beam breaks up into many spots, each of which is also approximately the spot size of the transmitter's diffraction limit.[16] Beam wander is a wavelength-independent phenomenon, but beam spreading exhibits a weak theoretical wavelength dependence.[16] Under conditions of severe turbulence, larger experimental values for beam wander can be obtained. Extensive experience with astronomical observations through the atmosphere is that the random behavior of the refractive index limits observations to a few seconds of arc.[15] Thus beam widths can be anticipated still to be small, with beam width increasing due to turbulence to the order of an arc second, equivalent to about 5 μrad, within the atmosphere.

Let us examine a beam directed vertically downward through the atmosphere. If we assume a divergence angle of 0.32 μrad at the source, then at the top of the atmosphere the diffraction-limited spot size will be approximately $0.32 \times 10^{-6} \times 1000 \times 10^5 = 32$ cm. While traversing the atmosphere, the beam size will increase due to turbulence. By using the figure of 5 μrad through a propagation distance of 10 km for the thickness of the atmosphere, one can obtain an estimate of a further widening of the beam spot by about $5 \times 10^{-6} \times 10 \times 10^5 = 5$ cm. Thus the diffraction-limited beam spot size on the ground will be increased by effects of turbulence in the atmosphere by about 15% to about 40 cm. The total area of the beam spot would thus be increased by atmospheric turbulence by about 50%, and the energy density incident on the target consequently reduced to about 45% of its previous value.

At high laser power levels, absorption of radiation can induce temperature changes in the atmosphere, which in turn result in density changes, and therefore index of refraction changes. These then alter the optical characteristics of the medium, and nonlinear phenomena, in particular thermal blooming, result.

Although thermal blooming in air appears to be able to set in as low as a threshold of a power level of about 1 kW, it is at high irradiances that thermal blooming becomes of interest.[15,16] A laser beam directed from a battle station satellite toward a surface target will enter the atmosphere only after traversing 1 000 km or more in space. Thus the energy fluxes will be of the order of a few kW/cm^2, fairly small compared to the hundreds of kilowatts per square centimeter characteristic of focal plane irradiances discussed in connection with thermal blooming.[16] So it would appear that nonlinear optical phenomena may have only minor effects on this case.

An additional effect on the energy density of the laser beam spot is, of course, the orientation of the target surface relative to the laser beam direction. Initially, normal incidence on a target surface was treated for conceptual simplicity. However, consideration of surface orientations at different angles to the laser beam will introduce angular corrections that will reduce the effective intensity by the cosine of the angle of incidence. While this is a minor correction, it is necessary for different zenith angles, and it may be of interest if specific types of targets (e.g., combustible roof structures or combustible vertical external walls or tinder materials on horizontal surfaces) were to be considered.

The efficacy of laser beams operating in incendiary attack mode from space through the atmosphere toward ground targets can thus be reduced by a number of effects. Furthermore, the efficacy of such weapons would be critically dependent upon weather conditions, primarily for efficient transmission of laser energy to surface targets, but also because surface weather conditions (e.g., heavy snow cover) can significantly modify the effectiveness of energy deposition at the target in causing incendiary effects. However, it should be noted that use in offensive rather than defensive mode is envisaged in this application, so that not only the most suitable targets, but also the time (and weather conditions) would be of the attacker's choosing. Under suitable conditions, it would appear that a full-scale incendiary attack from a large-scale system of orbiting laser battle stations or ground-based lasers with relay satellites might still pose an immense incendiary threat, possibly, as discussed, even marginally capable of initiating nuclear-winter-like climatic changes.

References

1. R. P. Turco, O. B. Toon, T. P. Ackerman, J. B. Pollack, and C. Sagan, Science **222**, 1283 (1983).
2. R. P. Turco, O. B. Toon, T. P. Ackerman, J. B. Pollack, and C. Sagan, Sci. Am. **251**, 33 (1984).
3. U. S. National Academy of Sciences; National Research Council, *The Effects on the Atmosphere of a Major Nuclear Exchange* (National Academy Press, Washington, DC, 1984).
4. G. Yonas, Phys. Today **38** (6), 24 (1985).
5. U.S. Department of Defense, *Report to Congress on the Strategic Defense Initiative* (U.S. Department of Defense, Washington, DC, 1985).
6. A. B. Carter, "Directed Energy Missile Defense in Space," background paper prepared under contract for the Office of Technology Assessment, Washington, DC, 1984.
7. A. L. Latter and E. A. Martinelli, "SDI: Defense or Retaliation?" *RDA Logicon*, R&D Associates, Marina Del Rey, CA, 28 May 1985.
8. K. Tsipis, *Arsenal: Understanding Weapons in the Nuclear Age* (Simon and Schuster, New York, 1983), pp. 183–211.
9. *The Effects of Nuclear Weapons*, 3rd ed. edited by S. Glasstone and P. J. Dolan (U.S. Departments of Defense and Energy, U.S. GPO, Washington, DC, 1977), pp. 282–300.
10. C. H. V. Ebert, Weatherwise **16**, 70 (1985).
11. R. L. Garwin, Nature **315**, 286 (1985).
12. R. A. McClatchey, R. W. Fenn, J. E. A. Selby, F. E. Volz, and J. S. Garing, in *Handbook of Optics*, edited by W. G. Driscoll and W. Vaughan (McGraw-Hill, New York, 1978).
13. H. S. Stewart and R. F. Hopfield, in *Applied Optics and Optical Engineering*, edited by R. Kingslake (Academic, New York, 1965), Vol. 1.
14. United States Air Force, *Handbook of Geophysics* (Macmillan, New York, 1960).
15. *Laser Beam Propagation in the Atmosphere*, edited by J. W. Strohbehn (Springer-Verlag, New York, 1978).
16. C. B. Hogge, in *High Energy Lasers and their Applications*, edited by S. Jacobs, M. Sargent III, and M. O. Scully (Addison-Wesley, Reading, MA, 1974).

Appendix F: Aids for teaching civil defense

compiled by L. C. Shepley

This appendix is written to help teachers incorporate topics important in the debate on civil defense (as well as many arms race topics) into courses they teach. This appendix will also be of use to those who participate in public forums on technical issues germane to civil defense.

Figure 1 delineates the educational activities of physicists. The aim of the nontechnical group of activities is to provide the basis for laypersons to obtain and to express informed opinions on issues pertaining to civil defense. A recent list[1] of 70 courses dealing with defense and nuclear war (most of which are outside the physics curriculum) gives some idea of the acceptance of these topics into today's curriculum. The technical activities involve teaching courses which are aimed in part at showing technical students how to use their skills to contribute to understanding public policy issues.[2] Figure 2 denotes physics subject areas typically taught in survey courses (as represented by Refs. 3–5). Table I contains a list of concepts and applications for each of these subject areas, with suggested references. Table II is a guide to many of the references, sorted by types. Table III is a guide to journals. The reference list at the end has brief annotations.

The physics of nuclear weapons and nuclear war already are subjects taught in many courses.[1] Within this general framework is civil defense, and we have listed those aspects of civil defense appropriate for courses or discussions led by physicists. Physicists also are often asked to use their expertise in public forums on nuclear topics, including civil defense. However, many of the problems with civil defense are ones of economics, psychology, sociology, or politics. For example, physicists will not have special insight on the alleged provocative nature of civil defense programs, though they may aid in the technical education of those who do.

Especially in the field of civil defense, even well-informed people may have differing opinions.[22] It is often the case that a purely physics course will rely on a single text, but in this area a variety of readings is desirable, and we have listed several sources. The educational role of physicists is a most important one. To be informed means to know, at least, some of the issues on which physics bears directly and to know the physical principles behind the issues. We feel that physicists should incorporate ideas relating to the nuclear arms race and to civil defense in courses and in talks where appropriate, and the purpose of this appendix is to assist them in doing so.

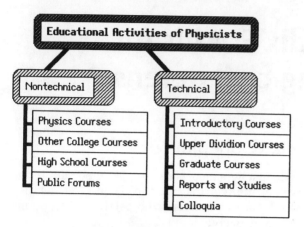

Figure 1. Educational activities of physicists.

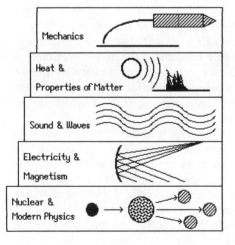

Figure 2. Physics subject areas in survey courses.

Table I. Concepts and applications for physics subject areas in survey courses.

Concepts	Applications
Mechanics	
Postion and velocity	Projectile motion (Ref. 5)
Acceleration	Missile trajectories and accuracies (Ref. 6) Traffic and evacuation plans (Ref. 7) Warning time (Refs. 8–10)
Force and pressure	Hardening of silos (Ref. 6) Blast and wind effects (Ref. 11)
Energy and momentum	Medical effects (Ref. 11)
Heat and properties of matter	
Generation of heat	Fireball and thermal wave (Ref. 11)
Propagation and effects of heat	Flammability; firestorms (Refs. 12 and 13)
Chemical reactions	Depletion of ozone layer (Ref. 14) Medical effects of burns (Ref. 11)
Heat budget of earth	Nuclear winter; properties of soot and dust (Refs. 15 and 16)
Sound and waves	
Propagation of waves	Blast waves and Mach front (Ref. 11)
Diffraction, reflection, and refraction	Dynamic and static overpressure (Ref. 11)
Seismic waves	Missile silo design (Ref. 6) Verification of test ban treaties (Ref. 10) Effects on humans (Ref. 11)
Electricity and Magnetism	
Induction and resistance	Electromagnetic pulse (EMP) (Refs. 11 and 17)
Forces on particles	Directed energy weapons (Ref. 16)
Lasers, including x rays	Strategic Defense Initiative (Ref. 16)
Nuclear and modern physics	
Nuclear reactions	Fission and fusion bombs; enhanced radiation weapons (Refs. 11 and 18)
Radioactive decay	Fallout (Ref. 11) Shelter design; monitoring (Ref. 13) Medical effects (Ref. 11)

Table II. Text references sorted by category.

	Text references Category

Refs. 6 and 8–10
Suitable as texts on nuclear war; quality of civil defense discussion varies. Some have teachers manuals.

Refs. 19 and 20
Articles aimed at physicists on a variety of topics.

Refs. 21–26
Popular descriptions, in a down-to-earth style. Sometimes imprecise in technical concepts, and sometimes containing views of a particular political group.

Refs. 27–33, 49, and 51
Policy discussions with a variety of viewpoints, some highly critical of civil defense programs.

Refs. 7 and 34–38
Source books with a variety of readings, mostly nontechnical.

Refs. 39–41, and 50
Manuals on survival techniques.

Refs. 12, 14, 15, and 42–47
Nuclear winter and other long-term consequences, including environmental, economic, and political effects.

Refs. 11, 13, 16–18
Technical aspects of weapons and their effects.

Refs. 1, 2, 10, and 48
Guides to the literature, including lists of sources on all aspects of the arms race, and lists of where to find other sources.

Table III. Journal guide.

Journals	Description
American Journal of Physics	Pedagogical aspects of physics, including relations to public policy
Armed Forces Journal	Military viewpoints
Arms Control Today	Newsletter of Arms Control Association
Aviation Week and Space Technology	Defense industry viewpoints
Bulletin of the Atomic Scientists	Science and public affairs journal, especially for arms control
The Defense Monitor	Newsletter for Center for Defense Information
Foreign Affairs	International relations journal
Foreign Policy	International relations journal
International Security	International relations journal
Journal of Civil Defense	Journal for advocates of civil defense
Journal of Defense and Diplomacy	International relations journal
Public Interest Report	Newsletter of Federation of American Scientists
New York Times Index	"Atomic weapons" and related categories
Nuclear Times	Peace movement journal
Nucleus	Newsletter of Union of Concerned Scientists
Oak Ridge Reports	Journal including various studies on defense issues.
Science	Particularly check each issue's News and Comment section
Scientific American	Popular science journal with articles and news notes on arms control
Science News	Weekly science news journal with news on arms control issues

References

1. UCAM (United Campuses to Prevent Nuclear War), *Summary of Courses on Nuclear War* (pamphlet, obtained from 1346 Connecticut Ave., NW, Washington, DC 20036) issued in 1984.
2. *Nuclear War: A Teaching Guide*, edited by D. Ringler (*Bulletin of the Atomic Scientists*, Chicago, IL, 1984). Especially see the article by L. G. Paldy, "Physical Sciences, Mathematics, and Engineering," pp. 7–8. Short articles on teaching about nuclear arms. Includes a list of subjects suitable for science courses and a very good guide to source material.
3. J. Boleman, *Physics: An Introduction* (Prentice-Hall, Englewood Cliffs, NJ, 1985). A standard, nontechnical physics text.
4. P. G. Hewitt, *Conceptual Physics*, 5th ed. (Little, Brown, & Co., Boston, 1985). A standard, nontechnical physics text.
5. D. Halliday and R. Resnick, *Physics*, 3rd ed. (Parts I and II, combined) (Wiley, New York, 1978). A standard, calculus-based, introductory text.
6. K. Tsipis, *Arsenal: Understanding Weapons in the Nuclear Age* (Simon and Schuster, New York, 1983). Technology of weapons and war; a bit uneven, with mostly conceptual discussion intermixed with some technical material. Over 60 pages of appendices, mostly simple calculations. A little material on civil defense.
7. *The Counterfeit Ark: Crisis Relocation for Nuclear War,* edited by J. Leaning and L. Keyes (Ballinger, Cambridge, Massachusetts, 1984). Readings against crisis relocation.
8. P. C. Craig and J. A. Jungerman: *Nuclear Arms Race—Technology and Society* (McGraw-Hill, New York, 1986). A text, including discussions of effects of weapons, the arms race, and economic and political consequences of war. Has questions, a few problems, and an Instructor's Manual.
9. R. Ehrlich, *Waging Nuclear Peace: The Technology and Politics of Nuclear Weapons* (State University of New York Press, Albany, 1985). Study questions at the ends of chapters. A considerable amount of material on civil defense.
10. D. Schroeer, *Science, Technology, and the Nuclear Arms Race* (Wiley, New York, 1984). Details on weapons systems and technological aspects. Lots of references, annotated. Good chapter on civil defense, discussing technology and some arguments. Excellent teacher's manual, including homework and exam questions, examples, more references, source materials.
11. S. Glasstone and P. J. Dolan, *The Effects of Nuclear Weapons*, 3rd ed. (U.S. Departments of Defense and Energy, Washington, DC, 1977). The standard reference on the effects of nuclear weapons. Comes with circular slide rule.
12. A. A. Broyles, Am. J. Phys. 53, 323 (1985). Formulas and calculations pertaining to smoke and fires in a nuclear war.
13. Federal Emergency Management Agency (FEMA): *FEMA Attack Environment Manual* (Federal Emergency Management Agency, Washington, DC, 1982) Note: This is an updated version, with few changes, of the DCPA (Defense Civil Preparedness Agency) Attack Environment Manual, June, 1973. Nine chapters, in separate pamphlets numbered CPG 2-A1 to CPG 2-A9, on all aspects of what goes on during a nuclear attack. Written in the form of "panels," with one descriptive page and one illustration page per panel, about 35 panels per chapter. Very carefully done and informative.
14. Office of Technology Assessment (OTA), Congress of the United States, *The Effects of Nuclear War* (Allanheld, Osmun and Co., Totowa, NJ, 1980). Detailed study of several scenarios; no nuclear winter. The chapter on civil defense describes U.S. and Soviet programs.
15. B. G. Levi and T. Rothman, "Nuclear Winter: A Review of the Key Factors," Center for Energy and Environmental Studies, Princeton University, Report No. CEES 197, 1985; Phys. Today 38 (No. 9), 58 (1985), a semitechnical discussion of the relevant factors.

16. Office of Technology Assessment (OTA), Congress of the United States, *Strategic Defenses* (Princeton University Press, Princeton, New Jersey, 1986), contains two excellent, objective reports: "Ballistic Missile Defense Technologies" and "Anti-Satellite Weapons, Countermeasures, and Arms Control."
17. Defense Nuclear Agency (DNA), *DNA EMP Course Study Guide (Modules I–X)* (U.S. GPO, Washington, DC, 1983), Report No. BDM/W-82-305-Tr. A course in the form of teaching modules.
18. T. B. Cochran, W. M. Arkin, and M. M. Hoenig, *Nuclear Weapons Databook. Vol. I: U.S. Nuclear Forces and Capabilities* (Ballinger, Cambridge, MA, 1984). First of a projected eight volume set; future volumes will include strategy issues. Lots of details about actual weapons and delivery systems. Nothing on civil defense. Lots of photographs and graphs.
19. D. W. Hafemeister, Am. J. Phys. **41**, 1191 (1973); **42**, 625 (1974); **44**, 86 (1976); **47**, 671 (1979); **48**, 112 (1980); **50**, 29 (1982); **50**, 713 (1982); **51**, 215 (1983). Science and Society Tests for Scientists, including tests on a broad range of defense and other issues. The questions include calculations of effects of nuclear war.
20. *Physics, Technology, and the Nuclear Arms Race* (AIP Conference Proceedings No. 104, American Institute of Physics, New York, 1983), articles and appendices on various topics. Proceedings of the Second Short Course on the Arms Race, held at the Baltimore Meeting of the American Physical Society, 17 April 1983, edited by D. W. Hafemeister and D. Schroeer.
21. B. Brodie and F. M. Brodie, *From Crossbow to H-Bomb* (revised and enlarged edition) (Indiana University Press, Bloomington, 1973). Little on nuclear war.
22. A. Carnesale, P. Doty, S. Hoffmann, S. P. Huntington, J. S. Nye, Jr., and S. D. Sagan (The Harvard Nuclear Study Group), *Living with Nuclear Weapons* (Bantam Books, New York, 1983). Very nontechnical discussion of issues. Very brief mention of civil defense.
23. Ground Zero, *Nuclear War: What's in It for You?* (Pocket Books, New York, 1982). A very readable, popular account of arsenals and possibilities.
24. New Manhattan Project, *Nuclear Mapping Kit* (American Friends Service Committee, 15 Rutherford Place, New York, NY 10003; 1979), a pamphlet.
25. G. H. Stine, *Confrontation in Space* (Prentice-Hall, Englewood Cliffs, NJ, 1981). Details, noncritical, on star wars systems. A bit dated.
26. *The Fallacy of Star Wars*, edited by J. Tirman (Union of Concerned Scientists) (Vintage Books, New York, 1984). A popular discussion by members of the Union of Concerned Scientists of star wars systems and faults.
27. L. R. Beres, *Apocalypse: Nuclear Catastrophe in World Politics* (University of Chicago Press, Chicago, 1980). Some civil defense.
28. J. Fallows, *National Defense* (Random House, New York, 1981). The U.S. defense establishment and policies.
29. D. Fischer, *Preventing War in the Nuclear Age* (Rowman and Allanheld, Totowa, NJ, 1984). Discusses the provocative nature of civil defense.
30. A. A. Jordan and W. J. Taylor, Jr., *American National Security: Policy and Process* (Johns Hopkins University Press, Baltimore, 1981). Discussions of policy, with a mention of civil defense. Discussion questions.
31. H. Kahn, *Thinking About the Unthinkable in the 1980s* (Simon and Schuster, New York, 1984). Chapter on civil defense, but nothing on nuclear winter.
32. R. J. Lifton and R. Falk, *Indefensible Weapons: The Political and Psychological Case Against Nuclearism* (Basic Books, New York, 1982). Critical of civil defense.
33. R. Scheer, *With Enough Shovels: Reagan, Bush and Nuclear War* (Random House, New York, 1982). Based on interviews, a critical account of administration policy makers, including civil defense attitudes.
34. *Search for Sanity: The Politics of Nuclear Weapons and Disarmament*, edited by

P. Joseph and S. Rosenblum (South End Press, Boston, 1984). Good source book for readings, though little on civil defense.

35. *The Nuclear Reader—Strategy, Weapons, War*, edited by C. W. Kegley, Jr. and E. R. Wittkopf (St. Martin's Press, New York, 1985). A collection of readings, from all sides, about issues. Thoughtful civil defense chapter by J. M. Weinstein.

36. *The Nuclear Crisis Reader*, edited by G. Prins (Vintage Books, New York, 1984). Good collection of articles, though not on civil defense.

37. *Toward Nuclear Disarmament and Global Security: A Search for Alternatives*, edited by B. H. Weston (Westview Press, Boulder, Colorado, 1984). Lots of articles, with questions for reflection and discussion.

38. *The American Atom: A Documentary History of Nuclear Policies from the Discovery of Fission to the Present, 1939–1984*, edited by R. C. Williams and P. L. Cantelon (University of Pennsylvania Press, Philadelphia, 1984). A source book of documents, mostly from the early years; civil defense, for example, appears in a 1954 document.

39. B. D. Clayton, *Life After Doomsday: A Survivalist Guide to Nuclear War and Other Major Disasters* (Dial, New York, 1980). A manual on survival, fairly well written.

40. R. L. Cruit and R. L. Cruit, *Survive the Coming Nuclear War: How to Do It* (Stein and Day, New York, 1984). A survivalist manual.

41. C. H. Kearney, *Nuclear War Survival Skills* (National Weather Service Research Bureau/Caroline House, Coos Bay, OR, 1982). One of the better survivalist books.

42. Committee on the Atmospheric Effects of Nuclear Explosions (G. F. Carrier, Chairman); Commission on Physical Sciences, Mathematics, and Resources; National Research Council, *The Effects on the Atmosphere of a Major Nuclear Exchange* (National Academy Press, Washington, DC, 1985). A good discussion on nuclear winter; includes bibliographies and index.

43. P. R. Ehrlich, C. Sagan, D. Kennedy, and W. O. Roberts, *The Cold and the Dark: The World after Nuclear War* (Norton, New York, 1984). An uncritical account of nuclear winter, based on a conference.

44. M. A. Harwell, *Nuclear Winter: The Human and Environmental Consequences of Nuclear War* (Springer-Verlag, New York, 1984). A 5000 Mt war is used as a base case and examined. Somewhat uncritical.

45. A. M. Katz, *Life After Nuclear War: The Economic and Social Impacts of Nuclear Attacks on the United States* (Ballinger, Cambridge, MA, 1982). Good specific studies, good illustrations, but no nuclear winter.

46. T. Powers, The Atlantic Monthly, **254** (No. 5), 53 (1984). A discussion of effects on policy of nuclear winter.

47. R. P. Turco, O. B. Toon, T. P. Ackerman, J. B. Pollack, and C. Sagan (TTAPS), Science **222**, 1283 (1983); Sci. Am. **251**, 33 (1984). One of the first and most influential studies of nuclear winter.

48. D. Schroeer and J. Dowling, Am. J. Phys. **50**, 786 (1982). An excellent, thorough listing of texts and articles, including descriptions of journals and papers on policies. Defensive systems, both passive (civil) and active are included.

49. F. J. Dyson, *Weapons and Hope* (Harper and Row, New York, 1984). An intelligent and thoughtful discussion of nuclear war issues.

50. J. Leaning, M. Leighton, J. Lamperti, and H. L. Abrams, *Programs for Surviving Nuclear War: A Critique* (published by the Bulletin of the Atomic Scientists, 5801 South Kenwood Ave., Chicago, Illinois 60637, 1983). An occasional paper featuring three articles on FEMA, crisis relocation, and the contingency hospital system.

51. E. Zuckerman, *The Day after World War III* (Viking, New York, 1984). Discusses government plans for surviving nuclear war.

Appendix G: Film bibliography

compiled by John Dowling

For a more detailed description of films on civil defense consult *1984 National Directory of Audiovisual Resources on Nuclear War and the Arms Race*, edited by Karen Sayer and John Dowling (Michigan Media, 400 Fourth St., Ann Arbor, MI 48103) (postpaid $4). This directory gives production information, ratings, and cheap rental sources for the films. An overall rating is given for each film, 1 is low and 10 is high. An update of this directory is maintained by John Dowling (Department of Physics, Mansfield University, Mansfield, PA 16933). This update lists new films released since the Directory was published and is available from Dowling for $3, postpaid.

About Fallout, produced by Office of Emergency Preparedness, Washington, DC [free loan from nearest Army Training and AV Support Center (write FEMA, Washington, DC 20472 for addresses); released 1963], 16 mm, color, 24 min. Illustrates the nature of fallout radiation, its effect on the cells of the body, what it would do to food and water after a nuclear attack, and what simple commonsense steps can be taken to guard against its dangers. Overall a 4.

The Atomic Cafe, directed by Kevin and Pierce Rafferty, Jayne Loader. (distributed by New Yorker Films, 16 W. 61st St., New York, NY 10023, videocassette by Direct Cinema Ltd., P.O. Box 69589, Los Angeles, CA 90069; released 1982), 16 mm, color, 88 min. Compilations of government civil defense films of the 1950s and 1960s which were produced to prepare the American public for a nuclear attack. Includes scenes such as Girl Scouts telling what foods to take into shelters, how to build shelters, duck and cover drills, what to do when the sirens go off, etc. Also featured are film and TV personalities and civil defense officials warning people of the dangers of nuclear war. Overall 7.

Briefly, About Fallout, produced by Office of Emergency Preparedness, Washington, DC [free loan from nearest Army Training & AV Support Center (write FEMA, Washington, DC 20472 for addresses); released 1967], 16 mm, color, 8 min. This is a shortened version of *About Fallout*. Overall 4.

Civil Defense Debate, produced and distributed by Disarmament Media Network of Oregon, 5111 SW View Point Terrace, Portland, OR 97201, released 1982, videocassette, color, 30 min. Portion of 1982 Civil Defense Council Convention in Portland. Panel discussion featuring Adm. Noel Gayler, Dr. Jennifer Leaning (PSR), and two civil defense advocates.

Eleven Steps to Survival, produced and distributed by National Film Board of Canada, 16th Floor, 1251 Ave. of the Americas, New York, NY 10020, released 1981, 16 mm and videocassette, color, animated, 21 min. Gives information and advice for survival in time of a nuclear war. Demonstrations on the construction of fallout shelters, the food, water, lighting, and diversions during the two-week waiting period are clearly outlined. Stresses the need for advance preparations and offers an address where viewers can write to obtain information on the material covered in the film. Overall 3.

For the Next 60 Seconds, produced and distributed by Serious Business Co., 1145 Madona Blvd., Oakland, CA 94610, released 1982, 16 mm, color, 3.5 min. Shows a man watching TV who sees some old weapons test footage. The civil defense tape goes on screen announcing a test of the emergency broadcast system. However... Overall 6.

The Front Line, produced and distributed by New Century Policies, P.O. Box 963, Boston, MA 02103, released 1982, videocassette, color, 28 min. Tells the story of how Cambridge, MA refused to treat safety for nuclear weapons the same way conventional public safety is treated. The tape includes local officials, citizens, and such experts as George Kistiakowsky, Helen Caldicott, and Randell Forsberg. Overall 4.

In the Event of Catastrophe, produced by WGBH for NOVA (distributed by King Features Entertainment, 235 E. 45th St., New York, NY 10017, released 1978), videocassette, color, 59 min. This NOVA production examines the question of whether there can be any defense against nuclear weapons. Includes interviews with proponents and opponents of civil defense. Lets both pro and con people have their say, but interposes no hard questions or judgments on either side. Overall 6.

Invisible Enemy: Fallout, produced by the U.S. Department of Army [free loan from nearest Army Training & AV Support Center (write FEMA, Washington, DC 20472 for addresses); released 1957], 16 mm, color, 28 min. Covers background radiation and heat, blast and radiation resulting from an uncontrolled nuclear explosion. Demonstrates fallout dangers and the construction and equipping of a fallout shelter.

The Last Epidemic, produced by Impact Productions (distributed by Resource Center for Nonviolence, P.O. Box 2324, Santa Cruz, CA 95063; released 1981), 16 mm or videocassette, color, 36 min. Records a 1980 conference of the Physicians for Social Responsibility. Consists of physicists, arms controllers, and physicians speaking on the general consequences of nuclear war: destruction of the ozone layer, ecological catastrophes, civil defense, and a detailed listing of the medical consequences involved in an attack. Overall 8.

Life After Doomsday, produced and distributed by ABC Video Enterprises, 1330 Avenue of the Americas, New York, NY 10019, released 1982, videocassette, color, 60 min. "Nightline" broadcast of a panel debate about civil defense preparation.

Memorandum to Industry, produced by Office of Emergency Preparedness, Washington, DC [free loan from nearest Army Training & AV Support Center (write FEMA, Washington, DC 20472 for addresses); released 1966], 16 mm, color, 8 min. Shows what industry can do for emergency preparedness: provide fallout protection and shelter for employees and the public, protect vital records, and assure continuity of company management in the event of an attack on the U.S. Overall 4.

No Place to Hide, produced by Lance Bird and Tom Johnson (distributed by Direct Cinema Ltd., P.O. Box 69589, Los Angeles, CA 90069; released 1982), 16 mm, color, 29 min. This chronicles the era of growing up in the shadow of the bomb. It covers civil defense procedures, fallout shelters, duck and cover exercises, etc., using vintage films, newsreels, cartoons, and TV programs that tried to sell the idea that nuclear war is survivable. The narration is strongly critical of government policies. Overall 7.

Nuclear Disaster, produced and distributed by Screenscope Inc., Suite 2000, 1022 Wilson Blvd., Arlington, VA 22209, released 1974, 16 mm, color, 14 min. Follows the activities of a film company making a documentary about nuclear disaster–has special surprise ending. Overall 7.

Nuclear War: A Guide to Armageddon, produced by Rich Green for BBC (distributed by Films Incorporated, 1144 Wilmette Ave., Wilmette, IL 60091; released 1982), 16 mm or videocassette, color, 30 min. Shows what would happen to London if a one megaton bomb were burst over St. Paul's Cathedral. It goes through all the blast, heat, and radiation effects at different locations for the blast. Its treatment of civil defense efforts is particularly devastating. Overall 7.

Nuclear War: Survive and Come Back Fighting, produced and distributed by Doomsday Studio, 1671 Argyle St., Halifax, Nova Scotia B3J 2B5 Canada, released 1981, 16 min, color, animation, 4 min. Satirical look at civil defense efforts done in animation. Overall 6.

Nuclear War: The Incurable Disease, produced by the International Physicians for the Prevention of Nuclear War. (distributed by Films Incorporated, 733 Green Bay Rd., Wilmette, IL 60091; released 1982), 16 mm and videocassette, color, 60 min. Three Soviet and three American doctors speak on the medical effects of nuclear war on prime time Soviet TV to over 100 000 000 Russian people. They discussed the death of 150 000 000 people, the collapse of the medical community, and the uselessness of civil defense if such a war occurs. Overall 7.

Nuclear War and You, produced by the Department of Defense (distributed by Nuclear War Graphics Project, 100 Nevada St., Northfield, MN 55057; released circa 1954), 16 mm, black and white, silent, 5 min. Shows effects of a nuclear blast on typical homes at various distances from a nuclear explosion. Overall 9.

Operation Cue, produced by Office of Emergency Preparedness, Washington, DC [free loan from nearest Army Training & AV Support Center (write FEMA, Washington, DC 20472 for addresses); released 1964], 16 mm, color, 14 min. Points out the contrast between the Nevada test in 1955 and modern thermonuclear devices, then continues as a documentary report on the Operation Cue exercise of 1955. The picture features unusual slow motion photography of the effects of blast on houses, radio towers, etc. Overall 4.

Protection in the Nuclear Age, produced by Trio Productions (distributed by the National AV Center, GSA, Washington, DC 20409; released 1980), 16 mm, color, 24 min. This is a somewhat overly optimistic estimate of the survivability of nuclear war. The narrator explains that "a nuclear attack is possible but we can survive. Nuclear war does not mean the end of the world or civilization." The film shows how to improvise fallout shelters in your basement and commonsense measures to take in the event of a nuclear attack. The film came under severe criticism and was withdrawn by FEMA, but it is still available from the NAV Center. Overall 3.

Public Safety Information, produced by Defense Civil Preparedness Agency, Washington, DC [free loan from nearest Army Training & AV Support Center (write FEMA, Washington, DC 20472 for addresses); released 1978], 16 mm, color, 14 min. A short emergency information film, quickly covering basic citizen information on protective actions to take in the event of enemy nuclear attack. Overall 4.

Radiation Effects on Farm Animals, produced by U.S. Department of Agriculture, Washington, DC (distributed by Film Library, University of South Carolina, Columbia, SC 29208; released 1964), 16 mm, color, 13 min. Provides information on immediate effects and long-term pathological changes caused by high radiation exposure. Discusses amount and rate of radiation exposure and the effects on the circulatory, digestive, respiratory and nervous systems of animals. Animation illustrates damage and regeneration of body cells.

Radioactive Fallout and Shelter (distributed by Educational Media Services, Boise State University, Boise, ID 83725; released 1965), 16 mm, color, 28 min. Teaches the individual how to take care of his medical and health needs in time of disaster when medical assistance might not readily be available.

Radiological Defense, (distributed by Center for Instruction Media & Technology, University of Connecticut, Storrs, CT 06268), 16 mm, black and white, 27 min. Shows dangers and problems connected with radioactive fallout and documents the status of radiological defense at the federal and local levels.

Outlines the effects of fallout on human, plant and animal life and the need for fallout shelters in case of nuclear war.

Safest Place, produced by the U.S. Department of Agriculture, Washington, DC (distributed by Center for Instruction Media & Technology, University of Connecticut, Storrs, CT 06268; released 1963), 16 mm, color, 13 min. Points out that farmers may decrease the possible danger from radioactive fallout resulting from nuclear attack. Describes relatively simple and inexpensive shelters which will give some protection for families and livestock.

Seattle After WW III, produced and distributed by KCTS-TV, 4045 Brookyn Ave. NE, Seattle, WA 91805, released 1982, videocassette, color, 30 min. Treats the effects of nuclear war on Seattle. Shows evacuation, shelters, crisis relocation, the physical and medical consequences of nuclear war on Seattle, and what life would be like after nuclear war. It covers a lot of material and does it fairly well. It gets "ordinary" people to talk about how they would handle such threats and how they would deal, or attempt to deal, with nuclear war. Overall 6.

Second to None, produced by ABC (distributed by MTI Teleprograms, Inc., 3710 Commercial Ave., Northbrook, IL 60062; released 1979), 16 mm and videocassette, color, 51 min. Reviews the origins of nuclear weaponry and witnesses a scenario for all-out nuclear war. Interviews with military and political leaders examine the elements of the debate over SALT II, the missile gap, U.S. preparedness, civil defense, the volunteer army, military intelligence, and options available to each superpower. This program gives an assessment of national security and the prospects for arms limitation. Provides a source of information for viewers concerned with political history, technology, civil defense, and the future of the armed forces. Overall 7.

Slanting, produced by Office of Emergency Preparedness [free loan from nearest Army Training & AV Support Center (write FEMA, Washington, DC 20472 for addresses); released 1964], 16 mm, color, 9 min. Uses animation and still photography to illustrate the "slanting" techniques used by architects to incorporate fallout shielding in buildings. Overall 4.

U.S. Security and Soviet Civil Defense, produced and distributed by Advanced International Studies Institute, 4330 East-West Highway, Suite 1122, Bethesda, MD 20014, released 1980, 16 mm and videocassette, color, 32 min. Discusses Soviet civil defense efforts and how they affect U.S. security. Assumes that nuclear war is in a sense winnable. Overall 4.

Under the Mushroom Cloud, produced and distributed by Nuclear War Graphics Project, 100 Nevada St., Northfield, MN 55057; released 1982, 35 mm sound and slide show, color, 21 min. Reviews civil defense of the 1950s and 1960s, discusses the resurgence of interest in civil defense, explains crisis relocation, critically examines its fundamental assumptions, and describes what nuclear war would really be like, and contrasts the options of preparation and prevention of war. Overall 6.

The War Game, produced by Peter Watkins for the British Film Institute (distributed by National Cinema Service, P.O. Box 43, Ho-Ho-Kus, NJ 07423; released 1968), 16 mm and videocassette, color, 49 min. The grim effects of a nuclear attack on Britain are shown in horrifying detail, based on information supplied by experts in nuclear defense, economics and medicine. Uses cinema verité style. Overall 10.

What About the Russians?, produced by Impact Productions (distributed by Educational Film and Video Project, 1725B Seabright Ave., Santa Cruz, Ca 95062); released 1983), 16 mm and videocassette, color, 26 min. Deals with such issues as "Are the Russians ahead in the nuclear arms race?" Why the concern over the MX, Cruise, and Pershing II missiles? Can we trust the Russians to honor a nuclear weapons treaty? What about Soviet civil defense? How can we end the nuclear weapons race and maintain our national security? Overall 7.

What Would Happen if a One Megaton Bomb Were Exploded on Lancaster, produced by the Lancaster Friends Meeting Youth Group (distributed by Wilmington College Peace Resource Center, Pyle Center, Box 1183, Wilmington, OH 45177; released 1982), 35 mm sound and slide set. This slide set was planned, photographed, and developed by the youth group, and they also wrote and taped the narration. The AFSC's *Nuclear Mapping Kit* was used in calculating probable deaths, injuries, and destruction. Other groups interested in undertaking a similar project may wish to see what this group produced.

Where Will You Hide?, produced by Audiographic Institute (distributed by University of Minnesota, Audio Visual Library Service, 3300 University Ave. SE, Minneapolis, MN 55414), 16 mm, color, 14 min. Shows that in the next war, none of us will be safe behind the guns of armed forces. Nuclear war will be directed at civilians and there will be no defense against nuclear weapons and nowhere to hide.

Index

Acroynms, xv–xvii
Agriculture, restoration of, 140–144
American Catholic Bishops, 2
Attack scenarios, 4–5, 81–82, 126

Berlin crisis, 22, 24–25
Blanchard, B. Wayne, 35, 44, 45
Blast shelters, 77–79, 82, 83, 87, 115–117, 171–197 (Appendix B), 199–202 (Appendix C)
Bomb shelters, 2, 6, 77
Breakaway, 49
Broadcast station protection plans, xv, 39, 43–44
BSPP (*see* Broadcast station protection plans)
Buildup factor, 67

Carter, Jimmy, 37
Chain reaction, 48
Chernobyl, 85, 119–122
Civil defense:
  aids for teaching, 215-222 (Appendix F)
  American attitudes towards, 11–32 (Chapter 2)
  appropriations, history of, 35
  arguments against, 203–206 (Appendix D)
  arguments for, 171–197 (Appendix B), 199–202 (Appendix C)
  benefits to society, 160–161
  Carter administration, 37
  costs to society, 119, 144–145, 160–161, 169
  criteria for effectiveness, 5–7
  Eisenhower administration, 36, 44
  evolution of, 34
  film bibliography, 223–228 (Appendix G)
  Ford administration, 37
  foreign countries, 105–124 (Chapter 8)
    Great Britain, 2–3, 11, 87
    Soviet Union, 21, 106–114, 194–196
    Switzerland, 77, 79, 81–82, 105, 114–119
  historical events pertinent to, 36
  history of U.S. civil defense, 35
  Johnson administration, 37
  Kennedy administration, 36–37
  lifeboat analogy, 99–100, 163–164
  long-term feasibility, 155–157
  Nixon administration, 37
  nuclear winter implications, 125–137 (Chapter 9)

Civil defense: (*Cont.*):
  overview of, 1–9 (Chapter 1)
  phases of defense, 3–4
  political issues in, 153–162 (Chapter 11)
  psychological issues in, 153–162 (Chapter 11)
  purposes of, 33–34
  Reagan administration, 37–38
  relative costs, 107, 119, 169
  setting for, 1–3
  short-term feasibility, 153–155
  strategic implications of, 95–100, 113, 157–160, 202
  Truman administration, 35-36
"Cognitive dissonance," 25
Compton effect, 67
Conflagrations, 52, 79, 209
Counter-force attack, ix, 4, 58–59
Counter-recovery attack, 58–59, 81
Counter-value, ix
Crisis relocation, 7–8, 37, 38, 40–42, 43, 85–104 (Chapter 7), 154–155, 158, 160, 192, 200–202, 203–206 (Appendix D)
  economic costs, 95, 200
  evolution of, 85–87
  nitty gritty of, 92–94
  rejection of, 87, 89, 115–116, 203–206 (Appendix D)
  spontaneous evacuation, 88, 201
  strategic implications of, 95–100, 113, 202
  underlying hopes and assumptions, 94–95
  what crisis relocation entails, 88–89
  would crisis relocation work, 91–92
Critical mass, 48
Cuban Missile Crisis, 24–27

Day After, The, 28
Direction and control programs, 43–44
Dog tags, 16
Doom Town, 15
Dosimeters, 72–73, 187
Double flash, 49
Downsizing, 66
Dual use, 37, 86, 89–91, 107, 112, 116
"Duck and cover," vii, 16

Eisenhower, Dwight D., 18, 36
  war games, 14–17
Electromagnetic pulse, xv, 43–44, 50, 56–58

230 Civil defense

Emergency Operation Centers, xv, 39, 43–44
Emergency operations planning, 39
Emergency planning zone, xv, 90
EMP (see Electromagnetic pulse)
EMZ (see Emergency planning zone)
EOC (see Emergency Operation Centers)
Essential worker protection, 39

Fallout, 51, 60, 65–75 (Chapter 5), 79–80, 82, 120–121, 141–143, 166, 167, 168
  patterns, 60, 79–80, 83, 166, 168, 189, 193
  protection factors, xvii, 60–61, 70–72, 77, 79–80, 83, 88–89, 167, 168, 190
  shelters, 25, 27, 73–75, 77–84 (Chapter 6), 87, 110, 115–117, 167, 168, 171–197 (Appendix B)
    contamination of, 80
    crisis of 1960–1962, 20–24
    development and surveys, 39, 41, 71
    dilemma of the shelter owner, 23
    expedient fallout shelters, 73–74, 80, 88
    radiological monitoring of, 42
    sheltering from a nuclear attack, 77–84 (Chapter 6), 167, 178–183
FCDA (see Federal Civil Defense Administration)
Federal Civil Defense Administration, 12–16
Federal Emergency Management Agency, viii, 27, 33–45 (Chapter 3)
  problems with FEMA programs, 40–44
  role and mandate, 33–34
FEMA (see Federal Emergency Management Agency)
Firestorms, 52–53, 62, 79, 185–187, 209–210
Ford, Gerald, 37
Fractionation, 69
Fratricide, 58

Gaither Report, 20
"Garrison-state mentality," 108
Glossary, xv–xvii
Gouré, Leon, 97, 106, 112
Graham, Lt. Gen. Daniel, 96
Gray, xvi, 176
GROB, 111, 194
Group fire, 52

Hiroshima, 19, 28, 52–53, 55, 148
Host areas, 37, 41, 88, 95, 98
Hurricane Carla, 92

IEMS (see Integrated Emergency Management System)
Industrial protection, 38, 39

Integrated Emergency Management System, 38–39, 86
Ivy League Exercise, 91

Johnson, Lyndon B., 37

Kennedy, John F., 21–23, 36
  Cuban Missile Crisis, 24–27
  fallout shelter crisis (1960–1962), 20–24
Khrushchev, Nikita, 20
Kyshtym, 121

Laser weapons, 208–213
Lucky Dragon, 18

National Security Decision Directive No. 26, 37–38
Nehnevajsa, Jiri, 26–27
Neutron bomb, 55
Nixon, Richard M., 37
Nuclear radiation, 65–75 (Chapter 5)
  alpha particles, xv, 65
  attenuation of, 67
  background on, 65–69
  beta particles, xv, 65
  clinical effects, 54–56, 66, 148–149, 166, 167, 168, 188
  exposure guidelines, 66
  gamma rays, xvi, 65
  initial nuclear radiation, xvi, 65–70
  monitoring of, 72–73
  neutrons, xvii, 65, 68
  prompt, 65
  radioactive decay, 65
  shielding, 67
Nuclear reactor emergencies, 89–91, 119–122
Nuclear war:
  effects of, 58–62
  images of, 17–20
  life after the bomb, 100–101
  long-range recovery from, 139–151 (Chapter 10)
Nuclear weapons
  air burst, 51, 57, 78
  blast effects, 53–55, 177–179, 184
  breakaway, 49
  chain reaction, 48
  crater from, 52–53
  critical mass, 48
  double flash, 49
  effective megatonnage, 60
  effects of, 47–63 (Chapter 4), 172–194
    single weapon, 50–58, 172–194
  fallout from (see Fallout)
  fireball, 49–50, 53, 60, 69
  fratricide, 58
  fusion weapons, 48–50

Nuclear weapons: *(Cont.)*
  ground burst, 51, 57
  physics of a one-megaton explosion, 47–50
  protection from, 171–197 (Appendix B), 203–206 (Appendix C)
  radiation effects, 54–56
  scaling, 58–60
  shock wave, 53–54
  strategic weapons, 47
  tactical weapons, 47
  theater weapons, 47
  thermal effects, 51–53, 55, 175–176, 178
Nuclear winter, 8, 28, 61–62, 101, 125–137 (Chapter 9), 143, 149, 201–202, 207–213 (Appendix E)
  baseline scenarios, 126
  current state of knowledge, 134
  implications for civil defense, 134–136
  model predictions, 125–131
  threshold, 127–128
  TTAPS, 125–131
  uncertainties in amount of smoke, 128–131
  uncertainties in climatic impacts, 131–134
  "year without a summer," 8, 134

Odum, Eugene P., 149
On the Beach, vii, 18
Operation Alert, 14, 18
Optical depth, 129–131
Ozone layer, 53, 61, 142–143

Pair production, 67
Peterson, Val, 15, 17–18, 44
Photoelectric effect, 67
Physicians for Social Responsibility, 27, 205
Plattsburgh evacuation, 92–94
Political and psychological issues in civil defense, 153–162 (Chapter 11)
  benefits to society, 160–161
  bibliography, 161
  costs to society, 160–161
  discussion questions, 162
  fear of attack, 13–16
  long-term feasibility, 155–157
  provocation, 157–160
  short-term feasibility, 153–155
Population protection (*see* Crisis relocation)
Proportionality, 2
Protection factors (*see* Fallout protection factors)

Psychological issues in civil defense (*see* Political and psychological issues)
Public opinion polls, 17–20, 24–28

Radioactive fallout, 19, 65–75 (Chapter 5)
Radioactivity, xvii, 16, 54–56, 60–61, 167
  attenuation of, 67
  clinical effects of, 54–56, 66, 148–149, 166, 167, 168, 188
Radiological defense, 37, 39, 42–43, 72–73
Radiological protection, 42–43, 72–73, 79–80
Rate meters, 72–73
Reagan, Ronald, 37–38
  seven-year proposed civil defense plan, 39–40
  Star Wars speech, 8–9, 86
Recovery from nuclear war, 139–151 (Chapter 10)
  agriculture, 140–144
  bibliography, 149
  discussion questions, 150
  economy, 144–146
  industry, 144–146
  livestock, 140–142
  national government, 146–149
  social order, 146–149
Risk areas, 37, 41, 88, 95, 98, 165
Rockefeller, Nelson, 21

SCOPE Report, 134–135
SDI (*see* Strategic Defense Initiative)
Seven–ten rule, 60, 166, 189–190
Shielding, 67
Spontaneous evacuation, 88, 201
Star Wars (*see* Strategic Defense Initiative)
Strategic Defense Initiative, 8–9, 101–102, 159–160, 207–213 (Appendix E)
Survival Town, 15
Survivalists, 28

Teller, Edward, 20, 21
Texas City Disaster, 94
Thermal blooming, 212
Three Mile Island, 91
Title V, 37, 39
Truman, Harry S., 20, 35–36
  Federal Civil Defense Administration, 12
TTAPS, 125–131

Wilson, Charles, 14–15
Wylie, Philip, 15, 17–18, 21